Epicureanism: A Very Short Introduction

T0054839

VERY SHORT INTRODUCTIONS are for anyone wanting a stimulating and accessible way into a new subject. They are written by experts, and have been translated into more than 40 different languages.

The Series began in 1995, and now covers a wide variety of topics in every discipline. The VSI library now contains over 450 volumes—a Very Short Introduction to everything from Psychology and Philosophy of Science to American History and Relativity—and continues to grow in every subject area.

Very Short Introductions available now:

ACCOUNTING Christopher Nobes
ADVERTISING Winston Fletcher
AFRICAN AMERICAN RELIGION
    Eddie S. Glaude Jr.
AFRICAN HISTORY John Parker and
    Richard Rathbone
AFRICAN RELIGIONS Jacob K. Olupona
AGNOSTICISM Robin Le Poidevin
ALEXANDER THE GREAT Hugh Bowden
ALGEBRA Peter M. Higgins
AMERICAN HISTORY Paul S. Boyer
AMERICAN IMMIGRATION
    David A. Gerber
AMERICAN LEGAL HISTORY
    G. Edward White
AMERICAN POLITICAL HISTORY
    Donald Critchlow
AMERICAN POLITICAL PARTIES
    AND ELECTIONS L. Sandy Maisel
AMERICAN POLITICS Richard M. Valelly
THE AMERICAN PRESIDENCY
    Charles O. Jones
THE AMERICAN REVOLUTION
    Robert J. Allison
AMERICAN SLAVERY
    Heather Andrea Williams
THE AMERICAN WEST Stephen Aron
AMERICAN WOMEN'S HISTORY
    Susan Ware
ANAESTHESIA Aidan O'Donnell
ANARCHISM Colin Ward
ANCIENT ASSYRIA Karen Radner
ANCIENT EGYPT Ian Shaw
ANCIENT EGYPTIAN ART AND
    ARCHITECTURE Christina Riggs

ANCIENT GREECE Paul Cartledge
THE ANCIENT NEAR EAST
    Amanda H. Podany
ANCIENT PHILOSOPHY Julia Annas
ANCIENT WARFARE Harry Sidebottom
ANGELS David Albert Jones
ANGLICANISM Mark Chapman
THE ANGLO-SAXON AGE John Blair
THE ANIMAL KINGDOM
    Peter Holland
ANIMAL RIGHTS David DeGrazia
THE ANTARCTIC Klaus Dodds
ANTISEMITISM Steven Beller
ANXIETY Daniel Freeman and
    Jason Freeman
THE APOCRYPHAL GOSPELS
    Paul Foster
ARCHAEOLOGY Paul Bahn
ARCHITECTURE Andrew Ballantyne
ARISTOCRACY William Doyle
ARISTOTLE Jonathan Barnes
ART HISTORY Dana Arnold
ART THEORY Cynthia Freeland
ASTROBIOLOGY David C. Catling
ATHEISM Julian Baggini
AUGUSTINE Henry Chadwick
AUSTRALIA Kenneth Morgan
AUTISM Uta Frith
THE AVANT GARDE David Cottington
THE AZTECS David Carrasco
BACTERIA Sebastian G. B. Amyes
BARTHES Jonathan Culler
THE BEATS David Sterritt
BEAUTY Roger Scruton
BESTSELLERS John Sutherland

Available soon:

For more information visit our website

www.oup.com/vsi/

Catherine Wilson

EPICUREANISM

A Very Short Introduction

OXFORD
UNIVERSITY PRESS

OXFORD

UNIVERSITY PRESS

Great Clarendon Street, Oxford, OX2 6DP,
United Kingdom

Oxford University Press is a department of the University of Oxford.
It furthers the University's objective of excellence in research, scholarship,
and education by publishing worldwide. Oxford is a registered trade mark of
Oxford University Press in the UK and in certain other countries

Published in the United States of America by Oxford University Press
198 Madison Avenue, New York, NY 10016, United States of America

British Library Cataloguing in Publication Data
Data available

Library of Congress Control Number: 2015945426

ISBN 978-0-19-968832-6

Printed and bound by
CPI Group (UK) Ltd, Croydon, CR0 4YY

To my children Eva and David

Contents

Acknowledgements

I am grateful to Eva Fisher and Yale Weiss who each read the entire manuscript in draft and suggested many corrections and improvements, as well as to the experts who set me straight on a number of points. Editorial advice has been exceptional, and any remaining errors of fact and misunderstandings are altogether my own.

List of illustrations

Chapter 1
Introduction

To the contemporary reader, the word 'Epicureanism' is likely to conjure up images of dainty dishes and fussy choices requiring a considerable outlay of funds. A recent newspaper article describes an enterprising developer's plan for an 'Epicurean Village...peddling high-concept foods from mod spaces' to be built in Paris' fashionable Marais district:

> 'This is where the butcher's shop will go,' said Mr. Naudon. 'Over there will be the cheesemonger, where the cheese will be hidden in designer drawers and taken out and explained.'... He pointed to other outlets designed for an organic bakery, an oyster bar, a fishmonger, ethnic restaurants, and a coterie of neighborhood mainstays... A team of international designers, including Tom Dixon and Maud Bury is styling 36 storefronts... [Mr. Naudon] never appears in public without a silk scarf and a Rolex.

Epicureanism, you might think, involves either marginal enhancements to the daily round or frivolity and wasted resources. Philosophy doesn't seem to enter into it.

For most of history, the term 'Epicurean' conjured up a very different image. Not only were stylish accessories never mentioned in this connection, even the association with fancy foodstuffs was infrequent. The Epicurean was often seen, however, as a depraved,

1. Bust of Epicurus.

adulterous libertine, usually drunk, a pig wallowing in the trough, and at the same time a threat to civil society. Dante located the Epicureans in the sixth circle of hell. Alternatively, Epicurus (Figure 1) was seen as a liberator, freeing humanity from manipulation, domination, and exploitation. In his *Utilitarianism* of 1863, John Stuart Mill noted the long-standing association between Epicureanism and swinishness, and commented that 'modern holders of the doctrine are occasionally made the subject of equally polite comparisons by German, French, and English assailants'. Mill himself held to the liberator conception of Epicureanism. He considered it the only sound philosophical basis for social reform.

Until recently, the philosophical and political significance of Epicureanism was not in doubt. But why was Epicureanism so divisive and controversial? What questions about the nature of reality, knowledge and experience, and the conduct of life did it propose to answer and how? How could it have been, and how might it still be, a compelling and attractive philosophy having nothing much to do with oyster bars and silk scarves?

The pursuit of pleasure was centrally important to the Epicurean system. As its founder is reported to have stated, 'I know not how to conceive the good, apart from the pleasures of taste, sexual pleasures, the pleasures of sound and the pleasures of beautiful form'. This statement signalled a bold break with the dominant schools of ancient philosophy, including those of Plato, the Stoics, and Aristotle, which insisted that the pleasures of the senses were different from and in conflict with the true good. Pleasure needed a defender, and Epicurus courageously and intelligently planted a flag.

Born in 341 BCE on the island of Samos in the eastern Aegean Sea, Epicurus studied philosophy in Asia Minor and taught in several Greek cities before moving definitively to Athens in 307 BCE. He remained the leader of his school in Athens until his death, probably from kidney stones, one of the most unpleasant and painful conditions known to medical science, at the age of 71 in 270 BCE.

Ancient Greek philosophy was organized into schools or sects each led by a teacher. Each had a location, or set of locations, where lecturing and philosophical discussion was conducted (Figure 2). Socrates held his sessions in the open air in the marketplace of Athens and Aristotle in a school building, the Lyceum. Epicurus held his sessions in his small house or in his private garden, a treed location (Greek gardens did not have many flowers) about a kilometre outside the city.

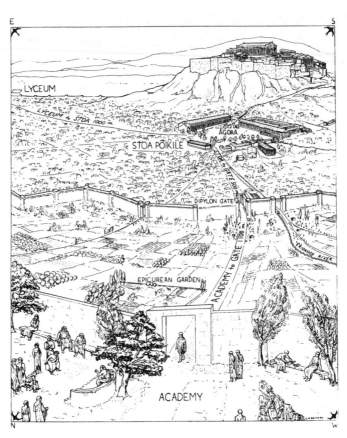

2. Map showing the approximate locations of the rival schools of Athenian philosophy.

Like the other philosophical systems of the ancient Greeks, Epicureanism embodied a comprehensive set of teachings about nature and its living inhabitants, and especially about human beings. It comprised a theory of experience and knowledge and an ethical and religious stance. It was the only system of antiquity that, without going so far as to declare that no god or gods existed, and that belief in divinities was a delusion, asserted that everything that happens is the work of nature and that nothing is to be feared or hoped for from any supernatural entity. The controversial but also attractive Epicurean prescriptions for the conduct of life and the attitudes to take towards love, friendship, death, and politics, were presented as following from fundamental truths about the constitution of the universe.

The ethical perspective of Epicurus was transformed and extended by the Roman poet Titus Carus Lucretius, a contemporary of Cicero and Vergil, at the same time as he vividly presented the Epicurean philosophy of nature. Lucretius' *De Rerum Natura (On the Nature of Things)*, composed in the middle of the 1st century BCE, had a considerable impact on science, anthropology, and moral theory when printed versions began to circulate in the second half of the 15th century.

Although Greek women in the 4th century BCE were beginning to enjoy greater personal freedom and economic rights, the Epicurean school was unusual for its time in that it had women from the class of *hetairai* as participants, so that the sexes mixed freely, taking their meals communally. The *hetairai*—the word is usually translated prejudicially as 'courtesans', with the emphasis on their sexual availability rather than their accomplishments, power, and status—were women who had either refused conventional marriage with its extreme restrictions or who, as foreigners, were unable to marry Athenian citizens. The *hetairai* were typically well read and conversant in the arts and philosophy, and they assumed the same privileges in their friendships and intimate relationships as men did, with one important difference. Although they were recognized as belonging to a different category from the ordinary (and inexpensive) prostitutes, the *hetairai* were financially dependent on male providers. As Chapter 7 will show, the division of women into three classes according to their social roles and status created difficulties for human relationships that affected and greatly interested philosophers.

Epicurus had, in any case, an affectionate personality, and his personal affairs together with these co-educational arrangements aroused much disparaging comment. Yet according to the 3rd century CE Greek historian of philosophy, Diogenes Laertius, who reported these stories, Epicurus was revered for his 'unsurpassed goodwill...his gratitude to parents, his generosity to his brothers, his gentleness to his servants', as well as—somewhat surprisingly—his 'piety towards the gods and...affection for his country'.

Though a prolific writer, Epicurus left few manuscripts to his immediate successors. For centuries, all that scholars possessed were texts and reports collected by Diogenes, including letters to Herodotus, Pythocles, and Menoeceus, and versions of the so-called 'Sovereign Maxims'. He is known however to have composed a collection of treatises on topics such as Love, Music, the Gods, Fate, Disease, and Kingship, as well as thirty-seven short volumes (papyrus scrolls) collectively titled *On Nature*. Once assumed lost forever, most of *On Nature* has recently been recovered in the excavation of Herculaneum, a centre of Roman Epicureanism destroyed by the eruption of Mount Vesuvius in 79 CE. The badly burned scrolls have been unrolled where this was possible, and restored and deciphered as far as possible, in a remarkable and painstaking process. Fortunately, *On Nature* was known to Lucretius, and his *On the Nature of Things* was based on Epicurus' original work.

Much of our predecessors' understanding of Epicurus came via his Roman critics, especially Marcus Tullius Cicero, a contemporary of Lucretius, and the moralist Lucius Mestrius Plutarch. Although Cicero was not himself a Stoic, his major works featured short expositions of Epicurean views and much longer expositions of anti-Epicurean Stoic views. While the two philosophies were in agreement that a tranquil life free of painful disturbances was best, they disagreed on every other matter of importance. The Stoics subscribed to a theory of intelligent design and believed in the providential supervision of the universe. They believed that matter was passive and inert and that a spiritual substance pervaded the entire universe, organizing it and endowing living beings with sense and movement, and human beings with the power of thought. They embraced determinism and recommended the acceptance of Fate. They distinguished sharply between the morally right and the merely expedient. They held that intellectually capable people had a duty to participate in politics, and that friendly relations could extend to people at the far corners of the earth.

This imposing, dignified set of doctrines furnished a formidable alternative to Epicureanism, especially when the observations and arguments of the Epicureans that accounted for their rejection of Stoic principles were unavailable. Epicurus' texts were mostly lost, as noted, in the volcanic eruption, and Lucretius' text survived into the Middle Ages in only a few copies. Did Christian librarians wilfully destroy manuscripts of the poem? The Fathers of the Early Church regarded Epicureanism as dangerous and corrupting, as did most medieval Jewish and Islamic theologians. Nevertheless, Epicurus' teachings on the theory of atoms found some sophisticated defenders and developers in the 9th-century Muslim philosophers of the *kalām*. Their discussions, also reported by the Jewish philosopher Maimonides, later served as a source for Western philosophers, including Thomas Aquinas, and may have contributed to the formation of the 'theomechanical' image of nature discussed in Chapter 3.

The manuscript hunters of the Renaissance were eager to locate the written relics of antiquity—its comedies, histories, and poetry, and its philosophy. In 1417, a surviving copy of *On the Nature of Things* was rediscovered in a monastery library by the intrepid Poggio Bracciolini, who had combed southern Europe in search of interesting material. For a long time, the manuscript remained in the hands of his copyist, but it was one of the very earliest books to be printed, with the first edition appearing in 1473. Edited, translated, and republished, it became one of the most important texts to fall into the hands of philosophers, poets, and students of nature of the 17th and 18th century.

The present volume is not a technical study of Epicurus' writings. Rather, it is an examination of how Epicurean ideas and arguments addressing science, morality, and politics were adopted and adapted over the course of the last few centuries, and of how Epicureanism may still be important in modern life.

Chapter 2
Atomic worlds

Atoms and void

The 'atom' is the fundamental element of reality in the Epicurean system. It is described in Epicurus' *Letter to Herodotus* as a solid body, possessed of shape, size, and weight, but colourless, odourless, tasteless, and far too small to be seen. According to the theory, there is a finite number of different shapes of atoms but an infinite number of each shape, and they move and congregate in a void of infinite extent. And that—atoms and void—is all there really is. The sun, moon, and stars were formed by processes of atomic accretion, as were the seas, mountains, and plains of our world.

Our world is, however, only one of an infinite number of worlds, or *cosmoi*, each of them with stars and an earth, and perhaps with other forms of human beings and animals, separated by 'intercosmic' void space. As the number of atoms is infinite, they cannot be used up, Epicurus thought, in a single world. And why should our world be unique, he asked, when most of the items to be found in nature—mountains, crows, rivers, oak trees—are found in the plural?

Epicurus was not the first atomist. The Greek philosopher Democritus, who died the year before Epicurus was born, anticipated many of his views, and Democritus' own teacher Leucippus, about whom very little is known, possibly did so as well. A slightly earlier tradition of atomism seems to have arisen in Indian philosophy. Only Epicurus, however, anchored a philosophical tradition of atomism with his school, and his views and reasoning are laid out more clearly and completely than those of his Greek predecessors.

How did the Epicureans come to believe in this spare ontology, and what is its relationship to our contemporary understanding of the physical world? Epicurus argued in the *Letter* as follows. It is evident to our senses that there are bodies and empty space around these bodies. Now, nothing can come from nothing. So, for there to be anything—a world of solid objects in space—there must be primary elements that constitute it. The objects of everyday experience, from a crumb of bread to a mountain range, are all destructible. If you want to destroy anything natural or man-made, an animal, or a rock, or a house, all you have to do is bash it to pieces, or cut it up, or pound it flat, and the thing is gone.

This goes for mountains, as well as for houses and animals. But if everything in the universe without exception were destructible, everything would sooner or later be destroyed, either by its own internal motions or by the impact of things around it, or by a combination of both. If there were no absolutely permanent entities, there would be no material for the formation of new entities to replace the old. Yet new objects with new qualities are constantly appearing. Plants spring out of the ground; hens produce eggs that turn into chickens. For a new thing to appear, its indestructible parts must have aggregated at some point.

We observe that the array of bodies changes, that the world of solid objects looks, tastes, sounds, smells, and feels different from moment to moment, day to day, and year to year. The primary elements must account for changes in the world while being themselves unchangeable. They must be unchangeable because all change results from the rearrangement or removal of the parts of a thing, or the addition of new parts to it. A simple, partless entity sitting in the void could not change.

Following this logic, Epicurus concluded that there must exist 'atoms' (literally, the *a-tomic* or 'uncuttable'). They must be of such a hardness and such a smallness that they can resist all blows and all attempts to divide them. They must have dimensions—be extended—if they are to make up the substance of the world, but their parts cannot be removed from them. There must also, crucially, be void space between the atoms so they can move, come together, and disperse.

The Epicureans allowed for four basic kinds of motion in the atom. First, there was the fall of atoms 'downwards' relative to worlds, not to the infinite space in which they were situated. Second, there was a rebound motion resulting from collisions between atoms. Third, the atoms 'oscillated' within objects when bumped into by their neighbours. Fourth, they sometimes 'swerved'. The swerve was a deviation from the basic downward path. It occurred frequently enough to cause the entanglement of many atoms; the result was a universe containing objects of a sufficient size to be experienced instead of a universe in which individual imperceptible atoms simply rained down. The atomic swerve, Epicurus thought, could supply the basis for free will, if only by providing a model of spontaneous, unpredictable, undetermined action.

Unlike many later atomists, and unlike his Stoic rivals, Epicurus rejected determinism not only in the physical realm but also in the realm of human agency. Many logicians maintained that if a disjunction—an 'or' statement—involving the future, such as 'either there will be a sea battle tomorrow or there will not be', was true today, as it obviously seems to be, one of its disjuncts must already be true today. The battle is accordingly fated to happen or fated to be averted.

For the Stoics, this was a welcome conclusion. All that happened was part of the divine plan for the good of the universe, including human choices and decisions, and events unfolded in a chain to reveal a providential order. Epicurus denied this, maintaining that the future was genuinely open and that human beings could make choices that affected the future. Take the disjunctive proposition 'either I shall go to the market tomorrow, or I shall not'. That seems undeniable. Is it then the case that I am fated to go to the market tomorrow or that I cannot go there? Epicurus thought not: I can decide either way when the time comes. Some modern logicians will insist that the disjunctive proposition itself is actually neither true nor false.

To back up the abstract argument that all the objects of everyday experience are composed of invisible material particles, Lucretius added numerous observations. He noted the force of wind and water when stirred into motion, and our ability to detect heat and odours, which, he claimed, depended on contact between our sense organs and something corporeal. Clothes hung out to dry lose their moisture bit by bit, and rings, statues, and pavements are worn away over the course of many years by pressure and friction. Light can pass through some hard transparent substances like slices of horn, but water cannot, suggesting that the atoms of light must be smaller than the atoms of horn and the atoms of water larger. The lingering scent of an herb rubbed between the fingers must be caused by invisible particles. Lucretius notes that even the tiniest insects must have internal parts below the threshold of visibility. Finally, he appealed to the common spectacle of dust motes dancing in a sunbeam to suggest that, while the motes were not atoms, they behaved like them, drifting, and colliding.

The atoms themselves, the Epicureans argued, cannot possess colour. If the atoms were coloured, compound bodies, according to Lucretius, would be any and all colours (or all mud-coloured, one might think). It is their groupings and positions, as they aggregate into solid objects and substances like water and wine, that brings colour into existence. The colours of things depend on the light in which they are seen and the position from which they are seen. This beauty and changeability of colours fascinated Lucretius, and he devoted many lines of verse to colour phenomena. 'Consider the iridescence imparted by sunlight to the plumage that rings and garlands the neck of the dove', he urges us, or the peacock's tail, 'glossed with red garnets', or seeming to 'blend green emeralds with blue lazuli'. A tranquil sea can look blue or grey; but the wind can whip it into a white froth. We can appreciate that the primary elements are colourless by noting how, when a thing is broken into pieces—when a fabric is torn into its constituent threads, for example—the colour disappears from them. Warmth, scent, and sound too are dependent on the arrangement and motion of atoms that do not possess them according to the theory.

Lucretius explains that perception can transform an active swarm of atoms into a quiet, motionless object with seemingly uniform properties, thanks to our inability optically to resolve them. A flock of individual sheep grazing on a hillside, with their lambs frisking and gambolling, appears from a distance as a motionless white blur; a distant army of men with bronze shields tramping about and assembling itself looks like a 'motionless pool of bright light upon the plain'. By means of such comparisons, it is easy to understand how atoms with only shape, size, and motion can give rise to a colourful, noisy, scented, tasty world of textured objects, some rough, some smooth, some sweet, some bitter. Hard objects, Lucretius speculates, must have closely compacted atoms. Liquids must be composed of smooth, round atoms that can glide over one another like a handful of poppy seeds that can be spilled in the same way as water. Sweet-tasting substances like milk and honey must be composed of other smooth atoms, whereas wormwood and other bitter substances tear open the little passages on our tongues and produce an unpleasant taste.

Lucretius noted that even individuals of the same type or species exhibit differences. Each nestling, each lamb, is different from its fellows—otherwise offspring and mothers could not recognize one another. Close examination shows that each grain of corn on a single ear is different from the others, as is each shell on the beach. Yet the number of atomic shapes is not infinite. If it was so, Lucretius argued, new sounds, tastes, and colours would always be coming into existence, melodies more exquisite than anything previously encountered, tastes more delicious, and colours more beautiful. So, as Epicurus thought, there is a limit to the number of types of atoms, but there is an infinite number of each type and so 'an abundant supply of material for everything'. Appealing to an alphabet analogy, Lucretius argued that, as a small number of letters can give rise to a vast variety of words with different meanings, so a finite number of types of atoms can give rise to the enormous variety of nature.

Epicurean ontology conspicuously excludes certain entities and relations. Neither incorporeal souls nor bodiless divinities exist, according to the atomist, and magic, understood as power over things and persons that does not require physical contact, is impossible. Effects that seem to imply action at a distance, such as contagion, in fact involve the flow of toxic atoms invisible to the naked eye. Further, as is indicated by the Epicurean analysis of colour, the objects and qualities our language denotes exist only by convention, that is to say, in virtue of human needs, capabilities, actions, and decisions. Colours and scents depend on our perceptual apparatus and environmental conditions and reflect a need to know and communicate to others information about the environment.

Extrapolating, one might say that objects such as beds—also slaves, daughters, and snowflakes—'exist' only in the sense that our form of life makes it necessary or interesting to construct and employ these categories, to designate certain 'solid objects' by these terms, and to make decisions according to perceived category membership. In the world of a worm, 'beds' do not exist, nor do 'daughters', nor does the quality of being coloured 'scarlet'—for worms do not have colour vision. All qualities, relations, and categories are thus fluctuating and changeable and relative to observers. A day might come when 'slave' has only a historical meaning. Less plausibly, it might come to pass that it no longer mattered in human life whether anyone is or ever was the 'daughter' of anyone else. Then the term would fall out of use and the concept might cease to be fully intelligible.

Doubt and denial

To many who considered it, atomism and the infinity of worlds thesis seemed deeply implausible. The pre-Socratic philosopher Parmenides argued against the existence of the void on the grounds that 'what was not' obviously could not be. As for the atoms themselves, no one had ever seen such things, so why should anyone believe they existed? Surely, if a piece of chalk or cheese is divided, the resulting pieces are simply smaller portions of chalk and cheese. Why shouldn't the division into smaller and smaller portions continue, in theory, forever, even if in practice we can no longer see what is there after a certain point, or find a knife blade thin and sharp enough to make yet another cut? What could make the purported atom uncuttable?

Aristotle, who had considered the atomic theory presented by Epicurus' predecessor Democritus, thought it worthy of respect, but ultimately wrongheaded. Taking a more physical and less linguistic stance than Parmenides, Aristotle surmised that the speed of a moving body would be infinite in void space where there was no resistance. The universe, he thought, must therefore be filled with matter—much of it invisible to our eyes—and all moving bodies must displace others, as a swimming fish displaces the ambient water. And how could there be anything extended that was not divisible? How could qualities—the scent and taste of cheese, the white colour of chalk—be generated from nothing, that is to say from material substance that was odourless, tasteless, and colourless?

Other critics, including the Academic Sceptic Cicero, ridiculed the atomic 'swerve'. Christian authors, following Platonists and Stoics, maintained that the order and beauty of the visible world could not have arisen from the motion of atoms as described by Epicurus. Surely, its useful plants and edible animals, its intelligent human beings, and its celestial objects with their regular pathways through the sky must be the work of a wise and benevolent deity who cared lovingly for his creation. All the main schools of philosophy, as well as the theologians, argued for this 'top-down' imposition of order and harmony by a governing Intelligence either above the world or permeating it. The latter idea is presented in Cicero's treatise *On the Nature of the Gods*, which eloquently describes a 'governing element', possessed of a higher form of sense and reason than the human, that created the world 'for the sake of gods and men' and that continues to 'hold the world in its embrace'.

The Epicurean thesis of the multiplicity of worlds and their ongoing creation and destruction seemed absurd to most ancient and medieval philosophers. Aristotle thought it obvious that our earth was unique and eternal and that it lay at the very centre of the universe. Hundreds of years later, in medieval Christian theology, other possible worlds might be allowed. But the existence of other co-occurring actual worlds, especially worlds with people in them, would seem to reduce the Christian drama of the Creation, the Fall of Adam, the mission of Christ to save the world, and the Second Coming to a curious set of happenings in one small corner of the many-world universe, raising alarming questions about God's interest in what terrestrials thought and how they behaved.

The revival of ancient atomism

Nevertheless, with the republication of Lucretius' poem, and the publication and dissemination of Diogenes Laertius' account of

Epicurus and his teachings in the mid-16th century, atomism began to find favour, the criticisms of theologians and philosophers notwithstanding.

The heliocentric Copernican system gained adherents in the late 16th and early 17th centuries, propelling the multiplicity of worlds thesis from a supposition based on nothing firmer than the supposed infinity of atoms to an obvious deduction. According to Copernicus, our sun was actually a star. Why then might not every star be the sun of some other inhabited world? The Dominican friar and philosopher Giordano Bruno was burned at the stake in Rome in 1600 for promulgating numerous heresies including this one. By 1644, René Descartes could defend a version of the multiple worlds theory in his *Principles of Philosophy*, and science fiction and popular philosophy took to the idea with relish.

Another important factor in the favourable reception of Epicureanism was the development of an alchemical literature dealing with the transformation of substances by heating, mixing, and distillation. Alchemy was devoted to the search for new medicines to prolong life, ease pain, and cure disease, as well as to the production of dyes, perfumes, and artificial gems, and the attempted transmutation, or at least transformation, of base into noble metals. Atomic theory could explain the dramatic changes in qualities and powers afforded by the mixing and heating of substances far better than the view that every different substance possessed its own irreducible set of qualities.

Further, atomism provided an argument from hope. If the qualities of substances depended on the arrangement of their smallest particles, and if 'mechanical' operations were able to produce new powers and properties, human experimenters ought to be able to find ways to produce important—and lucrative—transformations in matter. It would not be necessary actually to see and manipulate the atoms if trial and error along with good record keeping could establish correct procedures.

At the same time, the microscope, which came into use in the early 17th century, suggested to some practitioners that they might actually be able to see the tiny particles composing not only gold, lead, water, and other substances, but also the tiny particles they supposed were responsible for the mysterious processes, seemingly involving action at a distance, of magnetism and contagion.

Under the name of the 'corpuscularian', or 'mechanical', or simply the 'new' philosophy (to distinguish it from the 'old' Aristotelian theory of matter, form, and essential qualities), a version of Epicurean atomism was adopted by nearly every major figure of the 17th century's so-called Scientific Revolution. Its most prominent adherents agreed that 'solid' particles of a uniform matter underlay the different sorts of substances—gold, lead, milk, blood. Light was conceived as either a wave-like disturbance in a material medium or else as a stream of material particles emitted from luminous bodies.

The chemist Robert Boyle noted, in a Lucretian vein, the changes in the appearance of napped fabrics like velvet when viewed from one or another angle, and the remarkable changes that could be produced by the combination of chemical substances. Following Boyle, John Locke distinguished between the 'primary', qualities of matter (size, shape, motion, weight, and solidity) that he thought were independent of human perception, and the 'secondary' qualities (colours, scents, tastes, and sounds) that came into being only when an observer was present. The 'powers' of substances—such as that of the sun to bleach linen or melt wax—were assigned to the category of 'tertiary' qualities; the presence of these powers was detected by means of the changes they induced in the secondary qualities of objects.

Within this basic agreement shared by early modern philosophers, there were divisions, and also differences from ancient atomism. For one thing, the ontology of atoms, void, and motion seemed too restrictive to many enquirers. For Francis Bacon, atoms were real and important, but they were not the only active entities in the universe. 'Spirits', often conceptualized as ethereal fluids or as scents with active powers, were understood as permeating bodies and accounting for some of their various properties, including the growth and nutrition of plants and animals. For other philosophers, 'plastic natures' had formative powers that shaped plant and animal bodies. The Newtonian force of attraction and Francis Glisson's vital irritability, seemed to be required to explain the full range of phenomena, celestial and terrestrial.

The cohesion of substances presented a problem: why wasn't the universe full of fine atomic powder? Lucretius proposed that bonds, like those between dye particles and wool, or between glue and boards, are created, and that the atoms are 'held in union by...hooks and eyes' of some sort. Critics pointed out that hooks and eyes had to cohere as well. In time, the supposition that the pointed shape of some atoms explained their acid taste on the tongue, or the eel-like shape of others the liquidity of water, came to seem fanciful.

Moreover, Epicurean atomism in its original form conflicted with the understanding of God in the Abrahamic religions—Judaism, Christianity, and Islam—as omnipotent and omnipresent. How could there be an atom so hard God could not divide it? How could there be a region (the void) from which God was absent and in which he had created nothing? How could the universe and the number of atoms be infinite if God was the only infinite substance? And how could the universe be understood as ruled and regulated by God if the atoms were self-moving and congregated by chance?

One solution to the problem of divine supervision, favoured by some Islamic theologians, was to suppose that the atom was a purely passive entity, devoid of motive and other powers, divisible by God but not by any earthly force. On this view, nothing ever actually moved. Rather, God recreated the universe of particles at each instant in a new configuration, giving rise to the illusion of causal efficacy and motion in ordinary objects.

Another solution was to theorize God as the author of the Laws of Nature. One of the great conceptual innovations of early modern science was the discovery of the laws of motion, the mathematical regularities that characterize falling bodies and bodies colliding and rebounding. For inanimate bodies to behave in lawful, mathematically describable ways, it seemed that a divine Intelligence of supreme power must have decided exactly how they were to move and commanded them accordingly and in an irresistible fashion. So God was supposed to have established the Laws of Nature at the first moment of creation in such a way that all further events and processes, possibly excepting any that depended on human free will, would follow from them in just the way He foresaw and approved. Descartes adopted this line of argument, and Pierre Gassendi, the French philosopher and physicist, worked tirelessly from the 1620s to his death in 1655 to publicize, expatiate upon, and defend Epicurean atomism and to insist that it could be made compatible with Christian doctrine. Their writings, along with those of the mathematician-physicists Galileo Galilei and Christiaan Huygens, were a major influence on the philosophers associated with the English Royal Society founded in 1660, including Boyle, Newton, and Locke.

Some version of this 'theomechanical' image of nature as a collection of corpuscles obedient to divinely imposed laws of nature was widely accepted at the end of the 17th and throughout the 18th century. The principle of inertia, that bodies do not change their state of motion without being subject to some external force, was an important and fruitful postulate, which, at the same time, left a lot for God to do by way of creating and maintaining the universe. However, the notion that matter is essentially passive or inert was not an element of the original

Epicurean conception, which emphasized rather the spontaneous activity of the atoms, their multiple combinatorial possibilities, and the marvellous structures and properties that could emerge from their aggregation and interaction without any direction from above. The passivity thesis was rather a means by which corpuscularian theory could be accommodated to the Christian requirement of a Nature altogether subservient to a god and governed by immutable and perfectly reliable laws.

The revived theory of atoms came under serious pressure even at the height of its popularity during the Scientific Revolution. The microscope did not fulfil the hope of actually being able to see individual atoms and understand how their configurations were related to the qualities and powers of substances. Instead of minute corpuscles, the microscope revealed the enormous complexity, regularity, and intricacy of the bodies of insects and the parts of plants. It revealed that ordinary translucent pond water teemed with diminutive, active 'animalcules'.

Such observations suggested to philosophers like Gottfried Wilhelm Leibniz, writing in the 1680s and 1690s, that complexity in nature went 'all the way down', and that the inert, solid corpuscle was a fiction. Leibniz, whose interests extended to physics, metaphysics, microscopy, natural history, and logic, agreed with the atomists that the universe had, in a fundamental sense, to be composed of indivisible simple entities that were indestructible and active. But he balked at the notion that an extended atom could really be indivisible and indestructible. The ultimate constituents of the universe, he supposed, must be unextended, and so indestructible mind-like entities—atoms again, but with a difference.

Newton's research introduced new problems for atomism. Against the theory that colour arises from the arrangement of the superficial particles of a body in interaction with light, Newton showed that colours were already contained in a ray of white light, throwing the corpuscularian theory of colour into confusion. Newton accepted the Epicurean void, but he argued that it was pervaded by forces— the force of gravity, electrical and magnetic forces, and also certain short-range attractive and repulsive forces that, in lieu of hooks and eyes, explained the cohesion of substances and their solidity. He went on to speculate that perhaps the quantity of solid matter in the universe was only as much as would fill a nutshell, and that forces acting in a mostly void universe produced the phenomenon of solid, resistant matter.

Philosophers in the 18th century were fascinated by this seeming dematerialization of bodies, and the suggestion that matter and spirit were perhaps not opposed concepts. Just when it seemed that the scheme of atoms and void might be defunct, atoms came back into the picture in a surprise twist, as chemistry became an exact science. Chemistry now dealt with the ratios in which 'elements', with unique weights, that could never be transmuted by human processing into other elements, could combine. In John Dalton's textbook of 1830, the atoms are represented as round balls (Figure 3).

The atom in modern science

Contemporary physical science has retained some elements of the Epicurean image of nature, but the notion that solid, invisible, differently shaped particles form macroscopic objects by becoming entangled with one another thanks to their hook-like projections is gone forever. The atom of modern chemistry—an atom of gold or oxygen, or any other element from the periodic table—can be depended upon to combine in fixed ratios with atoms of other elements. But it lacks Epicurean indivisibility and indestructibility. Transmutation of the elements by human processing, which the alchemists hoped for, is thus possible, though difficult, and some transmutation occurs in nature through the process of radioactive decay.

ELEMENTS.

		W.ᵗ			W.ᵗ
⊙	Hydrogen	1	⊕	Strontian	46
⊖	Azote	5	✳	Barytes	68
●	Carbon	54	Ⓘ	Iron	50
○	Oxygen	7	Ⓩ	Zinc	56
⊗	Phosphorus	9	Ⓒ	Copper	56
⊕	Sulphur	13	Ⓛ	Lead	90
◐	Magnesia	20	Ⓢ	Silver	190
⊜	Lime	24	Ⓖ	Gold	190
⦿	Soda	28	Ⓟ	Platina	190
⦿	Potash	42	✽	Mercury	167

3. Dalton's list of atomic elements, which included atoms of azote, lime, soda, and potash.

The chemical atom is itself composed of 'subatomic particles' of which there are numerous types and subtypes, including quarks, electrons, and neutrinos. Quantum mechanics informs us that the subatomic particles are not solid entities with fixed sizes and locations. Rather, all subatomic particles exhibit wave characteristics. In a further departure from Epicurean theory, the location and velocity of subatomic particles can in some situations be assigned and even measured, but, according to Heisenberg's uncertainty principle, the position and velocity of a particle cannot be established at the same time.

A central Epicurean argument was that all change implies the rearrangement of the smaller parts of the changing entity, so that on this view, the transformation of a truly fundamental particle into another sort of fundamental particle would be impossible. As subatomic particles can change into other particles, an Epicurean would not regard them as really fundamental; other, *truly* fundamental entities must underlie them. One proposal—quantum field theory—regards fields as the fundamental entities. An individual electron is an excitation of the electron field, an individual quark is an excitation of a quark field, in a way analogous to the way in which a wave on the sea is an excitation of the surface of the sea. This remarkable idea helps to explain a feature of the world that the Epicureans—and indeed natural philosophers and physicists through the 19th century who accepted the atomic theory of matter—found it impossible to explain and had to take for granted. Why did the atoms have the particular shapes and sizes they did, and why there were so many of the exact same kind?

To conclude, the early modern rediscovery of the Epicurean tradition contributed significantly to the Scientific Revolution of the 17th century. Physical science has retained the idea that beneath the threshold of perception reality does not contain entities that have the same qualities and properties as the objects of ordinary experience. It is the interaction between the elements of that fundamental reality and our sensory apparatus that gives rise to a visible, tangible world. Since the mid-17th century, physicists, psychologists, and philosophers have been trying, with imperfect success, to understand exactly what this underlying reality is and how the experienced world arises from its interaction with the material brain that is equally a part of nature.

Chapter 3
Knowledge and understanding

The conquest of fear

Greek philosophers frequently contrasted the scientific understanding of processes observed in nature with myth and popular superstition, and with what they considered to be inappropriate references to the agency of the gods in popular religion.

Psychologists assure us that human beings have an instinctive tendency to ascribe phenomena they cannot explain in common-sense terms, especially those that have a bearing on their interests, to the intentional actions of invisible agents. Polytheism comes naturally to the human mind. Famines, plagues, earthquakes, floods, thunderstorms, volcanoes, and other dramatic events that arrive unpredictably and disrupt everyday life were taken as signs of the anger of the gods for human misdeeds and their awful power of punishment. For the Greeks and Romans, the anger of Zeus or Jupiter might be manifested in thunderbolts, the wrath of Neptune in storms at sea. Sexual passions that upset the lives of individuals were referred to Aphrodite, good harvests to the approval of the goddess Ceres. Sigmund Freud found evidence in the Bible that Jehovah was originally a wrathful volcano god. In many theologies, not only disasters and boons but everything of consequence or even of inconsequence that happens, is referred to the agency of demons, angels, or gods. The god of the Christian Bible is one who marks the sparrow's fall.

Even for philosophers who had no truck with such anthropomorphic fancies as divine anger and approval, it was difficult not to appeal to the agency of the gods, especially in connection with celestial phenomena. The regular rising and setting of the sun, its appearance at different parts of the zodiac in different seasons, the slow progression of the fixed stars, and the phases of the moon were so unlike any form of motion observed on earth that supernatural powers seemed to be at work.

Even Aristotle, whose philosophy accords great importance to the motive powers of nature, invoked divinities to explain celestial and meteorological phenomena. For Aristotle, the ultimate source of motion was a so-called Prime Mover. The circular motions of the planets, the sun, and the moon depended on a set of circumscribed solid but transparent spheres pushed by intelligent subordinate movers that carried these bodies around the earth. The motions of the heavenly bodies in turn caused the changes of season, alternations of heat and cold, rain and dry weather, which in turn caused growth, ripening, and reproduction.

Both Epicurus and Lucretius agreed on what might seem strange to the modern reader, namely that both the regularity and the occasional disruption of celestial motions produces fear in observers, a fear arising from thinking of the heavenly bodies as manifesting divine power and as issuing warnings and portents. For the contemporary observer, the reaction to a fine sunset, or an enormous harvest moon, or even to an eclipse, is not likely to be fear. It takes some effort to transport oneself back to a time when the sky could have generated such deep unease. Yet the fear of comets, whose paths cross the stars, persisted well into the 18th century and still persists in some quarters. According to a *New York Times* article published in 1997, '[t]he conviction that comets are cosmic emissaries warning of impending doom runs so deep in the collective psyche that it is embedded in the English language. The word "disaster" comes from the Latin *dis-astra*, meaning "against the stars."'

Humans tend to be anxious about currently salient features of their environment. We do not fear the heavens, nor do we connect their power to the terrible earthquakes, tsunamis, and hurricanes that kill thousands of people at a time. Yet many people live in an ongoing state of fear of chemicals, pollutants, and certain foods, reflecting the same fear of invisible agents as beset our ancestors. Fearfulness promotes vigilance and caution, which in many cases is justified: indeed, we ought to fear what is happening to our atmosphere and the powers that are harming us by corrupting it. Fear may also, however, reflect the mistaken belief that certain habits and performances can ensure our safety.

The anticipation of punishment and reward dispensed from above is perhaps a built-in arrangement in the human mind for maintaining an awareness of right and wrong. In order for interdependent yet crafty beings like ourselves to live together, it is helpful if believe that our oversights, as well as our devious, self-seeking plots and manoeuvres, will be witnessed and charged against us by an invisible, all-seeing judge. Further, our first thought in the case of any disaster is typically 'Who is responsible for this?' When no human agent is a likely candidate for blame, it is natural to posit an angry, punitive superhuman agent. If the gods are thought to be capricious, and touchy—insofar as often only a slight cause or no cause of their wrath is discernible—anxiety is increased. Constant efforts to placate them with sacrifices and ceremonies will follow. These will be perceived as successful when misfortunes let up for a time.

The Epicureans regarded the alleviation of suffering and especially the suffering produced by fear and anxiety as the most important contribution of philosophy to life. If all phenomena, even unusual phenomena, arise from the arrangement, motion, and effects of atoms, we cannot be impressed by the power of the gods, and there is no need to placate them or to try to forestall their wrath. Humans will still need to deal with the many and difficult effects of storms, famines, and plagues, but they will not experience the oppressive anticipation of punishment and the futile retroactive search for a previous crime when none was committed. They will not be enslaved to useless ceremonies performed in the hope of preventing new disasters.

The explanation of nature

The Epicureans accordingly proposed to explain all meteorological and celestial phenomena in physical terms, to free the gods not only from monitoring good and evil in the universe, but also, as Epicurus puts it, from all 'burdensome labour'. Epicurus stressed that the scientific explanation of the 'non-evident' involves a search for analogies from experience. Real understanding can only be attained by one who 'clings always to the phenomena and who is able to contemplate together [with it] what is similar to the phenomena'.

In his *Letter to Pythocles*, Epicurus sets himself the task of providing plausible explanations for clouds, rain, thunder, lightning, whirlwinds, earthquakes, snow, comets, ice, the ring around the moon, eclipses, and the light of the moon, by reference to the shape, arrangement, and motion of the entities and microentities involved, employing analogies from everyday experience. Lightning, for example, may arise from friction in the clouds which generates huge sparks, and snowflakes may be shaped by the 'pores' in clouds through which chilled water is pressed. Lucretius explains magnetism as arising from the impact of particles emitted from the magnet, which disperse the air particles so that a vacuum with suction is created.

A noteworthy feature of the programme of materialist explanation is Lucretius' particle-based account of common illnesses including swollen feet, toothache, and fever. 'Our earth and sky', he believes, 'contain enough harmful germs to allow a measureless amount of disease to be produced'. The world, he maintains, is stocked with elements beneficial to the lives of certain animals and also elements that are toxic 'on account of their different natures, their different structures, and the different shapes of their component atoms'. There are poisonous trees that kill anyone who sleeps under them, and lakes that cause the birds flying over them to drop from the sky. Fumes from the mines poison miners; these are all caused by the 'countless seeds' resident in the earth. Book VI of *The Nature of Things* concludes with an account of the plague of Athens caused by a deadly miasma of atomic poison that spread from Egypt to Greece. The notion that common diseases arise from tiny material particles outside the body that get into it, rather than from humoral disruptions within the body or malign influences of an unspecified sort, was floated but not really accepted in European medicine until the 19th century.

The Epicureans faced several problems with regard to their programme of explanation. Some explanations are the true, or the best explanations, but others, especially those concerned with meteorological phenomena, are only possible. Because the atoms themselves were declared to be permanently outside the range of sense perception, explanations citing them could never be directly verified by observation. No one can see the tiny pores in clouds that Epicurus conjectures are responsible for snow.

Moreover, a variety of incompatible explanations might all be equally plausible. A feature of Epicurus' methodology is that he never proposes to give the true explanation of a meteorological or astronomical phenomenon. He admits that he cannot do more than propose multiple possible explanations that are consistent with the atomic theory. Indeed, he is often concerned to supply more than one. Maybe the moon shines by her own light; maybe by reflected light. He professes to be unsure as to whether the sun is kindled and quenched on a daily basis or simply hidden by the earth, and he ascribes the 'turnings' of the sun and the moon to the pressure of the air or to an original circular motion inherent in them. The phenomena can come about in many and various ways, and one must not be too enamoured of 'the method of unique explanations' or 'groundlessly reject' all except one thereby 'desiring to understand what cannot be understood'.

The 17th-century Scientific Revolution

To be sure, when it came to explaining the risings and settings of the sun and moon and the rotation of the heavens, Epicurus' attempt to derive the phenomena from physical causes was painfully laconic. How could invisible microprocesses really generate such dramatic macroprocesses; how could thunder and volcanoes, let alone the phases of the moon, really be the effect of atomic behaviour? Yet for all its weaknesses, the Epicurean programme of explaining everything appealed powerfully to philosophers in the period of the Scientific Revolution, even when the fear-, apprehension-, and guilt-reducing motives for pursuing scientific explanations began to diminish.

In the new setting, however, the goals and presuppositions of scientific explanation were profoundly different. For Bacon and Descartes, the chief advocates of scientific research in the early 17th century, the aim of scientific understanding was ultimately the control of nature and the amelioration of human suffering by intervention in natural processes. 'Human knowledge and human power meet in one', Bacon declared. Controlling the stars or the weather was out of the question, but if the powers and qualities of all substances depended on the arrangement of subvisible corpuscles of a uniform sort of matter, then new 'forms', as Bacon called them, might be imposed on these substances by mechanical processes.

To succeed in actually rearranging the corpuscles to create gold and new medicines, to repair defective bodily mechanisms, and to combat contagious diseases, it would be necessary to fasten on the 'right' explanation, or at least an explanation that permitted prediction and control. From the multiple possible explanations of any phenomenon, those that could be successfully applied had to be singled out.

Descartes was driven by the ambition to understand, master, and redirect nature. He began as a diligent student of anatomy, hoping to understand how the bodily machine, in the absence of a soul, might explain animal life and behaviour, and he was assisted by an instrument not available to the ancient Epicureans: the magnifying lens, which had come into use in the early 17th century.

Although he made significant progress in understanding perception and emotion, Descartes never achieved his aim of making useful discoveries in scientific medicine. As this goal receded, he began to aim instead at offering satisfactory explanations, stated in terms of matter and motion, of as many phenomena as he had time to consider—not only snow, hail, the magnet, and so on, but also perception, memory, and generation, or reproduction. These explanations were broadly similar to those proposed by Epicurus in the *Letter to Pythocles*. Here Descartes ran directly up against the Epicurean point that a number of possible explanations can seemingly explain the phenomena, without it being possible decisively to establish the truth of one.

The problem of what is known as 'theory choice' in the absence of conclusive evidence was well known to the ancients, not only as it was presented in the form of the need to choose between moral philosophies or modes of living, but also between competing medical and astronomical hypotheses. Several rival schemes of planetary motion could 'save the appearances' to some extent, including Aristarchus' noted but widely rejected view that the earth revolved around the sun.

What has been termed the 'clock problem' now came into view. A clock or a watch can contain any one of a number of internal mechanisms. These mechanisms—gears, springs, crystals—can produce one and the same observable macroprocess, the sweep of the hands around the dial. A watch, however, can be opened up to inspect the internal mechanism to see exactly how it works, whereas 'opening up' the human body, whether dead or alive, did not reveal how growth or generation occurred, or even how the warmth of the mammalian body was produced. The availability of multiple conjectural explanations clashed with the ambitions of the 'experimental' philosophers. There was no way to 'open up' gold or lead to figure out how the latter might be converted into the former. Microscopes of the 17th century failed to reveal the atomic structure of metals and drugs, the existence of tiny particles of electricity and magnetism, and the hidden machinery of living things.

Disappointment ensued. We shall never know, the philosopher John Locke lamented, what arrangement of corpuscles makes gold yellow or rhubarb purge. We shall never know, David Hume lamented half a century later, why bread and milk are not nourishing for tigers.

The corpuscularian theory was, after all, a hypothesis, needing to be defended against its competitor theories, namely the Aristotelian ontology of bare matter and quality-endowing form, or the theory of the four elements, air, earth, fire, and water, or the alchemical ontology of salt, sulphur, and mercury.

Further, the doctrine that all natural phenomena were to be explained mechanically in terms of the physical action of parts on contiguous parts was hardly uncontroversial. Even looking aside from the problem of explaining thought and experience in materialistic terms, there seemed to the new breed of experimental philosophers to be more to the universe than particles in motion. Few thought to explain electricity, magnetism, gravity, light, fermentation, chemical reactions, and vitality in purely corpuscularian terms. Forces and 'active principles' demanded recognition, and the hope was that they could be studied as lawful and as mathematically describable.

Further, not all philosophers were as eager as Epicurus had been to free the gods or god from burdensome labour. Many indeed, including Isaac Newton, looked for arguments and evidence to establish the Christian God's ongoing involvement in maintaining the natural world. The fact that gravity, or the attraction between all masses in the universe that Newton was the first to describe in precise terms, did not lend itself to a corpuscularian explanation afforded him just such an argument. The difficulty of explaining how the mind and body affected one another, where the instincts of animals came from, and how the species maintained their constant forms from generation to generation offered many opportunities for early modern philosophers to invoke God's wisdom, benevolence, and power.

The invisible world

Both Locke and Hume were wrong in some respects. We can now, thanks to developments in zoological science and physiology, explain why tiger metabolism is suited to meat by reference to subvisible entities and processes inside tigers and chemical reactions involving molecules. We can explain how the malleability and sheen of a lump of gold follow from the arrangement of its atoms and how the laxative properties of rhubarb follow from its agency in our intestines.

To the ancient Epicureans, the boundary between the visible (the 'evident') and the invisible (the 'non-evident') was fairly sharp. The human eye could not distinguish anything much smaller than a dust mote—an aggregate of atoms. But the boundary has shifted constantly since the 17th century. Light microscopes were gradually improved, and, in the 20th century, new apparatus such as the electron microscope, and new techniques for the visualization of the very small, made their appearance. Our power over nature has been developed to a corresponding degree, to a level undreamed of by the ancients.

Nevertheless, the problem of hypotheses involving subvisible entities and their verification remains central, not only in the philosophy of science but also in the application of theory to practice. Even with good techniques and technologies for visualization, interpretation, inference, and extrapolation are required. Even with our best instruments and our experimental and analytical methods, we may be unable to decide, at least at present, between two or more plausible accounts of the underlying mechanism responsible for some phenomenon.

This uncertainty is common in ongoing medical research. How do arteries become clogged, molecule-by-molecule, with plaque, and how do brains develop the tangled proteins characteristic of Alzheimer's disease? Is the build-up of cholesterol the result of a superfluity of cholesterol particles in the blood, or is it the response of the body trying to repair damage to the arterial wall by plastering it with sticky particles? How does a tumour grow or a wound close up? Or how does the liver regenerate itself after injury? What is the cause of migraine headaches?

We believe that subvisible microevents and microprocesses give rise to these phenomena, but there remain multiple conflicting hypotheses. Where large-scale physical processes, such as those occurring now in the interior of stars, or that occurred billions of years ago when the universe came into existence are concerned, many uncertainties remain.

Most scientists maintain that there is always a fact of the matter about how events in nature unfold, that there is always a uniquely true explanation for every real phenomenon, and that we will get to know it if science continues to be practised long enough. Such

35

'realism' is not however, the only possible position. One might hold, with Epicurus, that processes and events that are not evident are not candidates for knowledge. For Epicurus, we can only have certain knowledge of what appears to us, and sense perception is the criterion of truth. In recent times, Bas van Fraassen has argued along similar lines that we should not *believe* theories that refer to invisible entities, processes, and events, though we may *accept* them as 'empirically adequate'—as useful for prediction and control.

How then did Epicurus' claim that the senses were the ultimate criterion of truth fit with his commitment to invisible atoms, and to a general theory of their qualities and motions and a cosmological theory based upon them? Epicurus did not discuss this question, as far as we know, and if he made an effort, as Aristotle had, to refute the views of the rival schools of philosophy, it has not so far turned up in his known writings. He must have recognized that the atomic philosophy was only one general account of nature, though supported by his *a priori* reasoning about the need for indestructible units and based on intelligent inferences from sense-experience, especially the phenomenon of gradual attrition.

In Chapter 5, I will consider Epicurus' theory of perception and his views on knowledge and truth more directly. Here I want only to call attention to his emphasis on the role of analogy and conjecture in scientific enquiry and his commitment to physical mechanisms in place of supernatural powers. The apparent paradox in Epicurus' treatment of knowledge—why be exclusively committed to the theory of subvisible atoms when other explanations of the phenomena are possible?—can be softened, if not resolved, by remembering that the theory of atoms *predicts* and *explains* their own inaccessibility to the senses.

Chapter 4
Living, loving, dying

What is life?

Living things, according to the Epicureans, are composed of the same material particles that compose stones, water, the stars, and all other such substances and objects.

In some respects, living and non-living things are alike. Both individual living entities, such as a rosebush, an elm tree, or a salamander, and individual non-living entities, such as a boulder, a snowflake, or a lake, are aggregates of entangled atoms. All such entities, along with entire worlds with seas and mountain ranges, come into existence gradually as their parts are built up, and all are dissolved in time into their constituent atoms. There is a 'fixed limit', as Lucretius puts it, to the duration and powers of every individual thing except the atom. The existence of every other individual has a beginning and an end.

In other ways, living and non-living things are very different from one another. Living individuals seem to grow from small seed-like entities and undergo the process we call 'dying', processes somewhat different from those of enlarging or diminishing on the part on the non-living.

Living things are structurally more complex than non-living things; they take in nourishment from the atoms in their environment and expand their boundaries up to a certain point. When they have reached a proper size, they begin to reproduce their kind, and, at some point after the age of reproduction, they begin to die, either from toxins and illnesses, or accidents, or simply through wasting away. In many animals, there are features even more striking than nutrition, growth, and reproduction: a sensitivity to pleasure and pain, internal sensations and emotions, the ability to move, think, and perceive, and to make decisions and act purposively.

This is not to say that there is a clear line of demarcation between the living and the non-living. For the ancients, mould and mildew were not clearly differentiated from rust; from our perspective, prions and viruses are also indeterminate. They can reproduce their kind and they employ resources from their hosts to do so, but they do not have life cycles as plants and animals do.

Epicurus states decisively in the *Letter to Herodotus* that the soul is corporeal and composed, like everything else, of invisible particles. He describes them as exceptionally fine; they are distributed over the entire body making it alive, sensitive, and aware, which they can only do when 'confined' within the frame of the body. Lucretius follows him closely. The lives of animals depend on 'seeds of wind and warm heat' or 'spirit' that 'abandons the veins and quits the bones' at death. The vitality of plants, insects, molluscs, and reptiles and other seemingly living but cold and motionless entities is not discussed. As reptiles breathe and move, they may possess at least an airy vital spirit, while plants are either not alive, strictly speaking (as some philosophers supposed as late as the 18th century), or else must have particles of vital spirit in their sap.

The ultimate origins of life

How then did living things come to inhabit the initially lifeless *cosmoi*? The ancient Epicureans did not find the appearance of life overly puzzling. No special explanation is offered of how atoms can combine to form plants and animals. Apparently they saw this as no more of a problem than the formation of other striking combinations—the sun, moon, and stars, or waterfalls, or snowflakes.

Indeed, ancient philosophers generally acknowledged the spontaneous production of life from lifeless matter, or matter of a different sort. Insects like flies appear to be bred in decaying matter, and parasites appear in the muscles and organs of animal bodies, seemingly having been engendered there. Mice and other vermin seem to appear out of nowhere, and Lucretius believed the worms that appear after rainstorms derive from mud. Plants transform particles of earth and water into leaves, stems, and branches; their matter is transmuted in turn into animal and human bodies that may be devoured by wild beasts and the 'strong-winged' birds, as Lucretius puts it. Nature is a unity:

> [A]ll of us are sprung from celestial seed; all are begotten by that same father, from whom mother earth, the giver of life, receives the limpid drops of moisture.... [She] gives birth to lustrous crops, exuberant trees, and the human race; she gives birth also to all the species of wild beasts, providing them with the sustenance that enables them to feed their bodies, lead a pleasant life, and reproduce their kind.

The life cycles of plants and animals indicate that the materials for life are ubiquitous and the renewal of the living world ongoing, even in the constant presence of death and dying.

Lucretius tends to prefer the term *semina rerum*, the 'seeds of things', or *primordia rerum*, the 'primitives of things', to the terms *atomus* or *corpusculum*. This suggests that in addition to endorsing the 'abstract' atom of Epicurus, possessed only of size, shape, and motion, he is advancing a somewhat different and more vitalistic—but still purely material—theory of subvisible

39

living seeds scattered throughout nature, awaiting the right conditions to grow and develop. Whether this was Epicurus' view as well may be revealed by the examination of the manuscripts of Herculaneum. In any case, Lucretius suggested that a younger, more fertile earth could have engendered larger animals in the far distant past, while the old and somewhat worn-out earth of the present day can engender only small animals and insects.

Lucretius saw the earth as first producing grasses and saplings, then birds, then as creating 'wombs' in the earth and giving birth to animals. After all, he points out, 'living creatures cannot have dropped from heaven, nor can terrestrial animals have emerged from the briny gulfs of the sea'. Not all such engendered animals, he observes, would have been able to nourish themselves and to reproduce their kind. They would have died without leaving offspring. Only those that had the right structure to survive, mate, and produce similar plants and animals would have established lineages.

The Epicurean treatment of the origins of the world in unguided atomic combination, and the coming to be and passing away of its living inhabitants, was strikingly different from the other traditions of pagan antiquity and from Christian teaching.

The notion that an intelligent architect—with the power to bring completed works into existence—planned and then created the universe, determining the number and form of plant and animal species that would populate the earth, takes various forms in the world's mythologies.

One version familiar to Western Europeans is a fusion of Platonic philosophy with the account of ancient Hebraic people. Platonic philosophy, as laid out in Plato's influential *Timaeus*, posits a 'demiurge' or builder who creates a world that is as perfect as possible. In the Platonic account, the divinity begins with Ideas of the forms of the various species, substances, and general kinds, and proceeds to bring into existence representatives of those Ideas. The Ideas of the general kinds—Horse, Human, Oak Tree, and also, Plato argued, the Ideas of Goodness, Truth, and Beauty—were then exemplified imperfectly in the creation of a material world.

The account in the Judeo-Christian Bible, the Book of Genesis, begins with the creation *ex nihilo*—from nothing—of heaven and earth by God and, within six days, his successive creation of plants, animals, and an original pair of humans, Adam and Eve, the alleged ancestors or 'first parents' of everyone on earth. The Islamic account in the Qur'an, which draws on it, is parallel in most respects. The creation of the world and humankind is believed to reflect one or more specific purposes of God—whether 'glory', or a wish to be known, or in order to bestow love, or in fulfilment of some more obscure plan.

Hundreds of millions of people have found these 'top-down' accounts of creation more credible than the 'bottom-up' account of the atomists with their falling, swerving, entangled particles. Over and over in the writings of Western theology and in popular philosophy, one finds repeated the view that the harmony of nature—the fact that all animals are outfitted with bodies, habits, and locations that enable them to find their food, coexist with other species, and perpetuate their kind—could not have come to pass without the foresight and creative power of a divinity. The fact that humans find in the plant, animal, and mineral worlds the materials they need—food, fuel, metal, drugs—and that the cycle of the seasons enables us to engage in agriculture and animal husbandry, suggests to us that the world was created especially for us. Cicero's treatise *On the Nature of the Gods* contains a rhetorically powerful, virtually irresistible presentation of this view, ascribed to the Stoic Chrysippus.

The Epicureans were the target of that work. With the recovery of Epicurean philosophy and the growth of knowledge of nature, mythic and anthropocentric accounts of the origins of the universe and its terrestrial inhabitants began to be seriously questioned. Creationism was not abandoned, but the planning, manufacturing, and maintaining roles of God were reduced, while still allowing Him some agency. Creation in six days and Adam and Eve as progenitors of all mankind were theses rejected by the most prominent philosophers of the 17th century, most notably by Descartes.

Descartes proposed that God had first created the universe as a block of matter whose parts could move relative to one another, and then laid down for all eternity the laws of motion directing the behaviour of all material bodies, from stars and planets to the tiny corpuscles of which the 'corporeal' world was composed. There His activity ended. Given a sufficient length of time, all possible forms would appear, persisting for longer or shorter intervals, among them the plants and animals of our world. These living machines would have structures and systems for nourishment, growth, and reproduction. God would need to do nothing after creating matter, motion, and the laws of nature, except to implant incorporeal souls into human beings as they were conceived.

Descartes presented his account as a useful fiction, and his critics accorded it the status of a fantasy, and a dangerous one at that. Yet soon enough, geological discoveries involving fossils of animals then unknown put new pressure on the Genesis account and the story of Noah's Ark. It was increasingly realized that the earth was many tens of thousands, perhaps many hundreds of thousands, or perhaps even thousands of millions of years old, and that it had undergone huge upheavals, changes of climate, and 'revolutions'. The notion that some species of animal, such as dogs, foxes, and wolves, might have descended from a common ancestor gained ground. Although Lucretius' image of full-grown mammals emerging from wombs in the earth was not regarded as credible, it was suggested by French and English philosophers a good century before Charles Darwin wrote, that, from a small number of original species, or even a single original life form, all the others might have arisen by some transformative process.

The Epicurean ontology nevertheless seems rather thin when it comes to understanding how the beauty and complexity of the world have taken shape, even allowing for very long timescales. While acknowledging that 'the old Epicurean hypothesis' was 'commonly, and I believe, justly, esteemed the most absurd system, that has yet been proposed', David Hume's spokesperson, Philo, in his posthumously published and scandalous *Dialogues Concerning Natural Religion*, wondered whether it 'might not be brought to bear a faint appearance of probability'. The slight emendation Hume proposed was to posit a finite, rather than, as Epicurus had, an infinite number of particles. 'A finite number of particles', Hume observed, 'is only susceptible of finite transpositions: and it must happen, in an eternal duration, that every possible order or position must be tried an infinite number of times'. It was left to Darwin to explain how variation and selection could produce new species, but the basic idea—that time, chance, and the forces of the environment could produce and alter living forms—was already in play.

How is self-assembly possible?

The gradual 'self-assembly' or 'epigenesis' of complex organisms whose parts work together to maintain life has always seemed hard to understand, since, for an organism to mature and function, it seems that all its parts must already be 'there'. The theory of the 'equivocal' generation of insects and vermin was dealt a serious blow in the period of the early modern Scientific Revolution when it was established that female insects, as well as female birds, women, and other mammals, produce eggs that are necessary for generating offspring. Soon thereafter the spermatozoa were discovered, and there was a rush to the anti-Epicurean theomechanical notion that God at the Creation had encapsulated all the generations in miniature, each inside its parent or parent-to-be, in either the female egg or the male 'seminal animalcules'. This theory was decisively discarded in the last quarter of the 18th century.

There were two fundamental Epicurean ideas that helped to bridge the apparent gap between non-existence and the richly populated world, and the gap between matter and life. They continue to have significance in thinking about the problem of self-assembly. One is the idea that some accidentally arising forms persist as individuals because they have stability-maintaining properties. The other is the idea that some collections of forms persist over time as species because they have reproduction-friendly qualities, while others perish.

The problem of the origins of life in the early universe and in ordinary reproduction becomes more manageable when we recognize that the particles that make up a living thing are not after all restricted to downward motions, collisions, vibrations, and unpredictable 'swerves', and that not all complexity arises out of chance entanglement.

We now know that there are forces of attraction and repulsion at the level of atomic chemistry. The chemical elements apart from the lightest—hydrogen—were formed in the furnaces of the early stars, and 'organic' molecules composed of carbon, hydrogen, nitrogen, phosphorus, and oxygen can form spontaneously in the interplanetary spaces. It is possible that the organic molecules present in all life forms were produced in the abundant water, methane, and ammonia that was present in our own primeval ocean under conditions of heat, acidity, and electrical excitation that no longer exist. Or they may have originated in some other part of the cosmos and been carried here by comets or meteorites.

For life as we know it to get started, molecules had to stick together in mutually sustaining patterns and become capable of self-replication. Further, self-replicators had to become capable of minute variations that did not affect their self-replicative capacity.

Not all combinations are durable, and only a few organism-size combinations of molecules have precisely the anatomical structure, physiology, and behaviour that enables them to reproduce.

Lucretius pointed out that some combinations of limbs and organs, though they might be formed by chance, were not viable. Either their parts could not function appropriately together or the composites were incapable of reproduction. He argued that centaurs—man-horse hybrids—could never have existed as a species at any time in history or prehistory, because the anatomy, passions, rate of sexual maturation, and habits of humans and horses are too different to permit successful reproduction in a hybrid.

The transition from protein molecules to unicellular organisms remains mysterious in contemporary science, and there are even people today who profess to be unpersuaded that life could have arisen naturally. What should we say to the monkeys-with-typewriters objection: that even allowing a few billion years for it to happen, it is overwhelmingly unlikely that the monkeys bashing on typewriters would produce the complete works of Shakespeare, and equally unlikely that 'blind' physical and chemical processes could produce such a richly populated and well-integrated world?

The answer to the objection is that this is the wrong way to look at the problem of the generation of complex forms, which can instead result from successive iterations of the retention of simpler forms. Instead of asking after the likelihood of a single monkey producing the works of Shakespeare, we need to ask, first, what is the probability that a monkey bashing on a typewriter would produce a word of English of four or more letters in, let's say, ten years of constant bashing? The likelihood is probably pretty good.

Suppose we were able to 'save' every word of English the monkey produced and allow the monkey to recombine the saved words into strings through more bashing around on some other sort of machine. How long would it be before we had a meaningful English sentence? Suppose we could now save all the sentences that could be written from all the saved words. How long would it be before we had a complete and savable Shakespearean scene? A complete ands savable play? A complete copy of the works of Shakespeare? A few billion years ought to suffice, especially with more monkeys on the job. Provided nature can save, by not eliminating, some simpler forms, they can later find ways to join up with others into more complex forms.

You might object that in this order-from-chaos scenario an intelligent being—not a monkey—is watching for words, sentences, and play sequences and plucking them out. But an intelligent selector is unnecessary. You need only imagine that, as the saved words are randomly combined with other words, those strings that do not form sentences when matched against a grammatical template are eliminated, and that, as sentences are randomly combined, those that do not form a portion of a Shakespeare play when matched against the works of Shakespeare are eliminated.

We do not yet possess a grammatical template or algorithm of the sort envisioned by Noam Chomsky that can sort all finite strings of words into sentences and non-sentences of English, but one is assuredly within reach. By analogy, strings of organic molecules, cells, blobs, and organisms are 'matched' against the conditions of existence, and some can be 'saved' to become parts of more complex, biologically operational structures.

The possibility of the piecemeal assembly of the complex from the simple just sketched goes a long way to answering the objection that the world is too rich in life, complicated, and harmoniously ordered to be the result of unintelligent mechanical processes and chance. The Epicureans may have grasped this idea, though we have no explicit account of their reasoning. They correctly surmised that vegetable life preceded animal life—since animals need vegetables to survive, though not vice versa—and that humans were latecomers on the planet.

They supposed, however, that the power of the earth to produce very large animals was greater in earlier times than in their own time, which, if we go back 2.5 billion years to the origins of life on our planet, is obviously untrue. If we go back only some hundreds of millions of years ago to the Jurassic era, however, they were right; earlier climactic conditions favoured the emergence, first of the dinosaurs, but later of gigantic insects, birds, and mammals, including dragonflies, snakes, birds, bears, and sloths of a size unknown today. It would be interesting to learn whether ancient naturalists had in fact come across the bones or imprints of such monsters, and drawn the correct conclusion from them.

Generation and renewal

Far more puzzling to ancient philosophers than the emergence of living things from non-living matter was the generation of like from like. The ability of birds, fish, and mammals to produce copies of themselves had no evident explanation. Why did peacocks produce only peacocks and foxes only foxes? Why were two and only two sexes needed? Why couldn't people reproduce like plants by simply budding or sowing viable seeds?

Rejecting the Aristotelian theory that males provided the 'form', the superior principle of the offspring, and females merely the 'matter', the lowly inferior principle, the Epicureans favoured an egalitarian epigenetic account. The seminal fluids of male and female mingle during intercourse; these fluids contain elements corresponding to traits and features of the ancestral stock and give rise to the new animal.

By the mid-19th century, long after the theomechanical theories of the pre-existence of complete miniature organisms had been decisively abandoned, Darwin and others reverted to a remarkably Epicurean theory—pangenesis—according to which 'particles', drawn from all parts of the body of both parents and lodged in the egg and the seminal animalcules, were mingled as the embryo formed and developed. Today we accept a version of 'preformation' that depends on the notion of information or directivity residing in the 'genes' and on a particulate structure to the units of inheritance, while adopting at the same time the epigenetic view that the building of the organism begins with an egg that divides into a clump of cells bearing no initial visual similarity with the organism it will become.

Neither Epicurus nor Lucretius had any idea why sex was necessary. It was only in the 20th century that it was accepted that, in more complex organisms, mating permits new genetic combinations that have survival advantages over clones basically identical to their parent, and that there are at the same time inefficiencies in requiring the participation of three or more sexes that outweigh the advantages of recombination. It isn't the case that if two parents are good, three or four must be better.

Nevertheless, Lucretius' philosophical poem pays marked attention to love and sexual passion between males and females as they manifest themselves in the life cycles of animals. He insists on the strong desire in females—not only women but also female birds and beasts—as well as in males to couple, thereby producing offspring, and he comments on the strong bonds between animal mothers and their offspring. At the start of *The Nature of Things* (Figure 4), he memorably invokes 'Venus, power of life', the 'delight of human beings and the gods' for whom 'the creative earth thrusts up fragrant flowers', who strikes 'seductive love into the heart of every creature… implanting in it the passionate urge to reproduce its kind'.

Departing from the currently known texts of Epicurus and coming into his element as a poet, Lucretius seems to conceive of amorous

T. LUCRETIUS CARUS

OF THE

NATURE of *THINGS*,

IN SIX BOOKS.

ILLUSTRATED with

Proper and Useful NOTES.

Adorned with COPPER-PLATES,

·Curiously ENGRAVED

By *GUERNIER*, and others.

Carmina sublimis *tunc sunt peritura* Lucretî
Exitio Terras cum dabit una Dies.　　OVID.

VOL. II.

LONDON:
Printed for DANIEL BROWNE, at the *Black Swan*
without *Temple-Bar*.

MDCCXLIII.

4. 'The sublime verse of Lucretius will not perish until the final day of the earth'. Title page to an 18th-century illustrated edition of Lucretius' *De Rerum Natura*.

desire as a force effectively counteracting the forces of annihilation that bring about destruction and death. In this regard, he appears to follow the ancient philosopher Empedocles, who supposed that Love and Strife were primary principles responsible for generation and destruction. Insofar as official Epicurean ontology recognizes neither goddesses active on earth nor active principles apart from motion, these themes do not belong to Epicurean philosophy as such. Yet the loves and desires of animals can be seen as one of those chance inventions of blind nature that have had the effect of stabilizing life and enabling the various species to continue.

The Lucretian view that amorous motives pervade nature, maintaining and renewing it and delighting all who experience them, is in any case contrary to the Christian interpretation of sexual passion as a deplorable result of the Fall. Human mortality and so the need for generation followed from Adam's sin, according to the Bible, and, as St Paul reluctantly concluded, 'it is better to marry than to burn [with unsatisfied lust]'. For the Fathers of the Early Church, especially Augustine and Tertullian, desire is vile lust encouraged by the Devil. Virginity for men and women is in many churches still regarded as the holiest of all states and as conferring a special relationship to God. Although some early Christian sects favoured affectionate relationships between men and women, such relationships were not codified into Christian moral doctrine, at least until the Protestant Reformation.

The Epicurean valorization of friendship between men and women, and the Lucretian valorization of passionate love as a force of nature at work throughout the animal kingdom and deserving philosophical attention and even a kind of respect, thus contrasted sharply with contemporary Greek and later Christian teaching. While the less punitive trends in sexual morals and customs of the late 17th and early 18th century might be difficult to trace directly to the recovery of Epicureanism, they were most certainly related to a breakdown, though not a permanent one, in Church authority and control, and a rejection of the Christian interpretation of sex. These trends are evident in the details of legal cases, in diaries and letters, and in the concerns presented in amatory fiction, poetry, and drama, beginning with Mme de Layfayette's *Princesse de Cleves* and the writings of Aphra Behn (incidentally an admirer of Lucretius).

For the Epicurean philosopher, generation and dying are symmetrical processes. Every perceptible entity—indeed, every entity except the atom—has a fixed term that ends with the dispersal of its constituent particles into the cosmic flux where they become material for the generation of new living and non-living entities. As Lucretius puts it:

> [N]o visible object ever suffers total destruction, since nature renews one thing from another, and does not sanction the birth of anything unless she receives the compensation of another's death.... So the aggregate of all things is constantly refreshed, and mortal creatures live by mutual exchange. Some species grow, others dwindle; at short intervals the generation of living things are replaced, and, like runners, pass on the torch of life from hand to hand.

Death is described as a peaceful sleep, induced by the dispersal of the seeds of spirit and the soul atoms: 'It is like the case of a wine whose bouquet has evaporated, or of a perfume whose exquisite scent has dispersed into the air'.

The Epicurean philosophy of mortality will be treated in Chapter 8. But first, in Chapter 5, we take a look at the materialist theory of mind.

Chapter 5
Material minds

The Epicurean soul

To give accounts of meteorological and celestial phenomena
without reference to the gods was one task for the Epicureans. To
give accounts of the origins of life on earth and the processes of
generation was another. Accounting for experience and thought in
materialistic terms was yet another.

As noted in Chapter 4, the Epicureans maintained that the soul,
like everything else, was material, composed of very small, light,
mobile atoms that pervaded the entirety of the living body of an
animal. Reportedly, Epicurus supposed these soul-particles to take
three forms, corresponding to heat, air, and wind, and providing
accordingly for bodily warmth, movement, and emotion. An even
lighter fourth type of soul-particle, corresponding to no nameable
element, was said to be responsible for awareness and for the ease
and speed of thought. In this regard, the Epicureans departed
both from the Platonic tradition that taught that the soul was
indivisible and incorporeal, and from the Aristotelian tradition
that described the soul as a 'form' joined to the 'matter' of the
body. The only incorporeal entity for the Epicureans was the void.

What reason is there to suppose the mind to be in any sense
'material' rather than an Aristotelian 'form'—a complement to

matter—or a separate non-material Platonic substance? Lucretius claims that the supposition that souls pre-exist, waiting for the growth of bodies to enter into and animate them, is absurd, especially if one thinks of 'the matings and births of wild beasts'. A mind requires an animal body and can no more exist outside one than trees can live in the sky, or clouds in the sea, or blood in timber. The mind, he observes, grows up with the body. The timidity of the deer, the ferocity of the lion, and the rationality of human beings are fitted to their bodily forms and are not found in other sorts of bodies. The different admixtures of the four types of particles can further explain the temperaments and capabilities of the various species.

Mentality is accordingly explained by the activity of 'very small seeds, which form a chain throughout the veins, flesh, and sinews'. They are smooth and round, requiring only the lightest touch to set them in motion; they can flow like water and with great speed on that account. The notion that soul and spirit atoms pervade the human body, rendering it alive and sensitive, able to make decisions, and to move as it wishes to, has a firm basis in intuition. When a foot or an arm 'falls asleep', it becomes numb, and one can easily imagine the soul atoms flowing back into it as it awakens with a sensation of pins and needles. In a blackout or faint, the soul atoms seem to withdraw from the limbs or to become inactive, and certain drugs and alcohol seem to alter their state of excitation or repose. The notion that the soul is distributed and corporeal is in many ways more appealing than the notion that an incorporeal soul is lodged in the brain where it somehow directs the movements of a puppet-like body by a kind of telekinesis, and reads its bodily changes as sensations and perceptions.

Lucretius was definite in his opinion that the soul atoms were not individually alive and aware; they were not miniature persons. The individual atoms composing a human being cannot 'shake and tremble with uncontrollable laughter' or 'discourse at length on the structure of compounds', or investigate their own nature.

Rather, we can 'laugh without being formed of laughing atoms...
and...possess intelligence and expound philosophy...without
being composed of intelligent and eloquent seeds'. The 'things
that we see endowed with sensation [are] composed of seeds
that are absolutely devoid of sensation'. But the atoms have to be
'united and combined' in a particular way in order to confer
consciousness and feelings of pain and pleasure on an animal,
along with perceptions.

Perceiving, thinking, dreaming

Visual perception typically involves experiencing and getting to
know about objects and occurrences located at some distance
from the perceiver. Epicurus and Lucretius seem to have felt
themselves to be on firmer ground where the atomic analysis
of this process was concerned, than in trying to explain either
astronomical phenomena or generation. They did not cite
merely possible explanations, but the explanation they took to
be uniquely correct. It rested on a theory of *idola*, material
images emitted from objects that possessed their shapes and
colours, yet were so thin as to travel unperceived through the
void and stable enough to resist dissipation. Unlike certain
contemporary materialists, they did not equate perceptions,
dreams, and sensations with brain-states. Their theory of
material films projected in all directions by all objects was
somewhat curious, but it had considerable explanatory power
in the Epicurean system.

How is it possible in an atomic universe to see the distant stars or
even to observe the approach of another person some way off?
Epicurus explains that particles are 'continually streaming
off from the surface of bodies' although they are replenished so
quickly that we notice no diminution. Smell and hearing are
accomplished by the emission of particles, and so is vision. The
atoms, recall, are themselves colourless. Yet the atoms emitted
from bodies include certain ultrathin 'outlines' that 'share the

colour and shape of the objects' from which they originate, and that move so quickly that they 'give the presentation of a single continuous thing'. Lucretius amplifies this account by suggesting that the coloured shadows cast by awnings and banners of 'saffron, russet, and violet' are composed of atoms discharged by them, as scents, smoke, and heat are discharged from other objects.

Our imaginations are stimulated, Lucretius goes on to argue, by the 'countless subtle images of things [that] roam about in countless ways in all directions on every side'. This chaos of roving images accounts for the incoherence of dreams; when we are asleep, these films enter our bodies and so our minds in a jumbled fashion. In dreams, 'we fancy that we pass over sky and sea, rivers and mountains, and traverse plains on foot'. We can also visualize objects that do not exist in the waking world when images are combined. Hence we dream of centaurs that never were, and of the dead, whose images cannot populate the air when their flesh is decaying underground (Figure 5).

The activity of dream-persons is explained by the succession of images in different postures that blur into continuous movement. But how is it that we are not continuously assaulted by involuntary images, and that we can sometimes direct our own visual thoughts? Lucretius explains that we pay attention to only some out of the immense store, just as we focus on only a selection of the visual field.

Truth and error in perceptual experience

Epicurus' epistemology takes sense perception to be both the standard of meaning and the standard of truth. We need to grasp what is denoted by our words, he says, by tying them down to their perceptible referents rather than merely to definitions. Names are given to things with a particular sort of outline, or a characteristic appearance. Thus, it seems, only things we can point to have names, and things that do not have a particular sort of outline or

5. Hypnos and Thanatos, the gods of dreams and death.

a characteristic appearance that can be indicated to a learner cannot be the subjects of true statements. So much for centaurs and gods, but alas, this criterion would seem to exclude many abstract entities such as numbers and governments, which do not have outlines or appearances and are in fact understood through definitions, practices, and institutions. Epicurus seems to acknowledge this in allowing that our beliefs about various entities are derived from perception via analogy, resemblance, composition, and 'some slight aid from reasoning'. The existence and qualities of the atoms themselves, as noted in Chapter 3, must be just such an inference.

Both Epicurus and Lucretius recognize that, even if perception is the standard of truth, we are subject to certain optical illusions because of material interference with the transmission of the films. Square towers look round from a distance because the *idola* are battered, buffeted, and blunted on their journey. We can create double images by pressing against an eyeball, and a twin row of columns of equal height viewed from one end creates the impression that they taper to a point. When on board a ship, we may feel that we are motionless and that the hills and plains are flying by us, and when regarding the heavens we do not perceive the movement of the sun and moon.

Lucretius denies that these common visual phenomena ought to undermine our confidence in our senses. The fault, he says, lies not in the senses but in the 'reasoning of the mind'. This claim puts him in something of a quandary, for he seems to refer truth and falsity to judgements, while at the same time maintaining that 'our conception of truth is derived ultimately from the senses', which cannot be refuted.

Here he follows Epicurus in arguing that when there are two perceptions of the same modality, say two visual experiences, one cannot correct the other. The same holds of two perceptions of different modalities (say touch and vision); why would one have authority over the other? And reasoning, says Epicurus, cannot contradict sense; what reasoning would justify the inference that a distant tower that for all the world looks round is actually square? Lucretius realizes that he has fallen into a muddle, insofar as he admits that there are illusions and that they are the fault in some way of the mind, but he insists that 'if you were not prepared to trust the senses...life itself would at once collapse'.

The answer to questions about what is 'really' there turns out to involve reference to previously seen shapes and to favourable viewing conditions. Epicurus explains that although the senses rather than reason are the arbiters of truth, in case of conflicts and uncertainties—e.g. about whether one is seeing a horse or a cow, or whether a tower is round or square—one ought to defer judgement until one has examined the thing at close quarters. When he maintains that the hallucinations of madmen and the dreams of sleepers are 'true', he means only that they really occur.

Epicureanism and empiricism

The Epicurean theory of experience and judgement has many echoes in later philosophy, especially in the branch known as empiricism.

The notion that we form and reason with 'ideas' of objects only by combining sensory ideas of their qualities and powers implies that knowledge of the world is unattainable by intuition and revelation. Further, it implies that the mind is apt to be led astray by its facility in concocting complex ideas of non-existent objects. Linguistic terms, Bacon and Locke insisted, were meaningless unless tied to perceptions; Locke allowed only perception and reflection on one's own perceptions as sources of knowledge, and Hume maintained that all ideas were derived from previous 'impressions'. This claim furnished the basis of Hume's withering critique of metaphysics.

Empiricism showed up traditional philosophical terminology, especially that of medieval scholastic philosophy, in a bad light. Too many terms, its critics maintained, had been introduced into philosophical discourse that purported to refer to actual definite entities but that were vague, ambiguous, or vacuous. To prevent pointless disputation and needless quarrels, and to make progress in human knowledge, it was essential to purge language of all terms that could not be tacked down to perceptual experience. In the mid-20th century, 'ordinary language philosophy', encouraged by the philosophers John Austin and Ludwig Wittgenstein, was a renewed attempt to eliminate a recondite technical vocabulary from philosophy, and to bring philosophizing down to earth through a thoughtful consideration of how distinctions in ordinary language were related to actual experiences.

The success of these movements was incomplete. Towards the end of the 18th century, Kant and his followers insisted that empiricism was too limited a philosophy to address the moral and political problems of humankind. The Epicurean position that meaningful names must refer to entities that can be pointed to or that have 'outlines' continues to be contested by philosophers who argue that reality is not exhausted by such things. Mathematical entities and properties, for example, are assumed by mathematical Platonists to be as real as tables, chairs, and colours. The natural and human sciences as they are practised today depend on specialist vocabularies that contain words for unseen entities that no one can point to, such as 'population pressures' or 'the prevailing rate of interest'. The particular jargon of philosophy depends heavily on definitions and on contested or variable definitions. Such terms might or might not be accepted under Epicurus' allowance for the extension of perception by analogy, resemblance, and composition, and a minimal degree of reasoning.

The Epicurean insistence that our beliefs about the physical world ought to be recognized as true or false by being tested against experience—or evidence or data, as we might now say—is nevertheless widely accepted in the sciences. The other central feature of Epicurus' perception-based epistemology—his claim that all appearances are true because the senses cannot contradict themselves or one another, or be overruled by reason—remains problematic. We do seem to distinguish between 'veridical' perception and 'illusion'; and between perceptions, hallucinations, and dreams, regarding the last two as delusory or at least as presenting unreal scenes to the mind.

On closer analysis, however, it is clear that there is no firm basis for distinguishing between veridical and illusory perceptions except in an Epicurean manner, by reference to preferred perceptual standpoints. The moon looks larger at the horizon than overhead, but which appearance of the moon is an illusion? It is only because we see the moon less frequently on the horizon than overhead that we describe the great size of the rising or setting moon as an illusory appearance. We call an 'illusion' a perception that surprises or startles us or that is not confirmed by touch, as in the case of the Müller–Lyer illusion, involving two lines of equal length but with inward- or outward-pointing arrows on their ends. Alternatively, an illusion is a perception that would lead to a senseless or faulty plan, such as the decision to throw away a perfectly good spoon because it appears bent in water.

The mechanisms involved in illusions are, however, no different than those involved in perceptions we consider to be revealing the world to us just as it is. Where hallucinations and dreams are concerned, the surprises are extreme and any plans formulated on the basis of what appears will be very faulty indeed. There is nevertheless a basis for agreeing with the Epicureans that appearances cannot be false and simply are what they are.

The material mind and its critics

The theory of the material mind was embraced, or at least regarded as viable, by a number of early modern philosophers who were persuaded by Epicurean observations and considerations. Thomas Hobbes and Pierre Gassendi were convinced that the heart and brain were the seat of thought and feeling, and Locke suggested, somewhat timidly but influentially, that God might have given 'suitably organized' matter, such as was to be found in the human brain, the power of consciousness and thought.

Descartes, famously, resisted this conclusion. Despite his acceptance of a basically Epicurean theory of the physical world and the nature of scientific explanation, Descartes firmly denied in his *Meditations* that the soul was anything like a 'wind or a flame' pervading the animal body, thereby hitting out directly at the Epicurean theory of the material mind. 'Animal spirits', he conceded, did flow through the nerves and muscles, and they were altogether material, but their function was purely mechanical; they transmitted impulses to the brain or inflated and deflated the muscles. Souls played no role in maintaining life for Descartes.

But, in addition to matter, the created universe contained an array of human minds, attached to living, mechanical human bodies.

In constructing a new theory of visual perception to fit with this new conception of the person, Descartes scornfully rejected the Epicureans' 'little images flitting though the air'. Instead of material *idola* interacting with a material mind, Descartes insisted that the mind was an incorporeal, unextended substance that could nevertheless interact with the material body. This interaction took place in the very centre of the symmetrical-appearing brain, where the pineal gland, which Descartes believed to be a uniquely human structure, was located. When it communicated with the body in this way, the soul was able to experience, understand language, and initiate motion, which no purely material entity could do.

The rest of nature, Descartes argued, was just 'corporeal substance'—unconscious, blind, and mechanical. Animal bodies were material machines that moved in response to physical stimuli. Only the living human body was attached to a mind, enabling it to have feelings, emotions, sensations, perceptions, imaginings, dreams, thoughts, beliefs, memories, and 'volitions' or acts of will.

Paying the price for escaping from the theory of soul atoms and coloured films, Descartes introduced a problem of how physical events could cause or otherwise produce experiences in an incorporeal substance. It was debatable as to which theory was more implausible: coloured films flitting through the air, or an incorporeal mind that produced conscious experiences of coloured objects when lodged in a machine.

The Epicurean material mind is fully permeable to matter. The Epicurean depicts us as inhabiting an atmosphere of floating, flitting images, some of which we register as belonging to a world outside the mind, others of which we tag as memories, anticipations, and fantasies, or recall as dreams and hallucinations. We are in touch with slices of the world that necessarily resemble it.

The Cartesian model, by contrast, introduces the problem known as the 'veil of perception'. When the right physical stimulus and the right brain state are present, our minds produce an image of the environment with colours, odours, sounds, tastes, and textures, which we experience as 'in' the 'external' objects we see, smell, hear, taste, and feel, and also believe to be 'caused' by them. But if the world is, as the Cartesian model proposes, composed of Epicurean corpuscles, in what sense do we perceive and come to know about the 'real world'? Are we not aware only of our own experiences and veiled off from the physical reality that we suppose produces them?

This problem was troubling to early modern metaphysicians. George Berkeley, an Anglican bishop who was seriously worried by the encroachment of Epicureanism in philosophy, was able to turn the problem of the veil into an attack on the very existence of 'matter', which he argued to be an incoherent notion. We perceive only qualities, he argued, including colours, shapes, sounds, and textures. These qualities are observer-dependent and changeable, and so can exist only within minds. Matter, which is allegedly external to all minds and unchangeable, therefore cannot be perceived.

What we call 'seeing a ripe cherry', according to Berkeley, is being aware of a small round red shape that we anticipate to taste sweet. Atoms are unperceivable and accordingly imaginary. They could not in any case have shape and size but be colourless, insofar as anything that is extended has to have some colour. The supposition that a 'material' coach that is colourless and soundless produces the experience of the noisy, colourful coach is accordingly absurd. We experience only our experiences, Berkeley argued, not a world external to all minds, and only the ongoing perceptions of an omnipotent and benevolent god make the order and continuity of human perceptions possible. In place of atoms and the void, Berkeley proposed that only minds, ideas, and a very active and necessary god are real.

Berkeley's arguments identified the major weaknesses of corpuscularian and especially Cartesian accounts of experience. They were cleverly contrived, but his conclusions conflicted with common sense, and in later life he went over to a more sensible view accepting the existence of unperceived corpuscles. Kant intervened in this dilemma to argue that Berkeleyan idealism had to be wrong. We had to suppose ourselves affected by an external world independent of our minds and God's mind, as the materialists took for granted.

Yet, for Kant, ultimate reality could not be atomic in case an atom was an even *possibly* perceivable object. The things ultimately responsible for our perceptual experience could not themselves be perceivable objects. They could not have shapes, sizes, and observable motions. Indeed, the 'thing in itself' or the 'things in themselves' responsible for our experiences could not be said literally to 'cause' our experiences, insofar as causality was a relationship between one perceivable event and another. Lucretius thought that the individual atoms, too small to be seen individually, collectively 'blurred' into the impression of continuous, uniform substances, but Kant denied that we perceived things in themselves in a blurry fashion by standing in a causal relationship to them.

Three philosophical alternatives to materialism that were worked out in the 17th and 18th centuries are dualism, panpsychism, and nescience. They all have their contemporary followers.

The dualist, following Descartes, maintains that the mind is an incorporeal substance that is related to a particular human or animal body. Some, though not all, dualists maintain that the mind can act upon the body and be acted upon by it, and that it can perhaps even survive the physical death of the body. Panpsychism, a view often ascribed to Spinoza, supposes consciousness to be everywhere in nature, in pebbles, raindrops, and atoms. It has a few, but only a few, modern defenders.

The third alternative is Locke's and Kant's nescience. It is the view that we can never explain the way in which perception and consciousness are related to an external world because our categories are equipped only to deal with causal relationships between perceptible things. Contemporary 'mysterians' who believe we will never have an acceptable theory of consciousness, that the mind cannot understand itself, are the inheritors of this position.

Materialists—in their contemporary instantiations—reject all of these views.

Materialism today

Although several hundred years of research on the nervous system have elapsed since Descartes took up the problem of the relationship of experience to the brain, it remains mysterious how physical and chemical entities such as neurons, synapses, ions, and neurotransmitters make a world appear to us and enable us to think, plan, experience emotions, act and react, do mathematics and science, and create works of art. To what extent, one might wonder, is materialism a live option, and how might contemporary materialism differ from the Epicurean account of the material soul and its workings?

The three alternatives to materialism just cited suffer from serious weaknesses. Causal interaction between an incorporeal soul and a corporeal body is impossible to conceptualize on any model of causal interaction we possess, and no one has any idea what a soul, released by death from its former body, could do or experience. Panpsychism was already rejected as ridiculous by Lucretius; it cannot explain why a collection of individual atomic perceptions should combine into an organism's individual, personal experiences. Kantian nescience seems an exaggerated response to a difficulty and a deterrent to future research. Perhaps philosophy cannot make any headway on the mind–body problem as long as it employs its traditional methods of making up imaginary scenarios and finding faults in abstract arguments, but for philosophy to pronounce *a priori* on what the empirical sciences can accomplish in this regard seems pretentious.

In considering the problem of how mind, brain, and world are related, we should accept that first, insofar as particles of 'solid matter', with their shapes, sizes, and weights, no longer play the role in our physics of fundamental entities, materialism, in the sense in which the Epicureans and their early modern successors understood it, is obsolete.

Further, no contemporary investigator will accept 'soul atoms' diffused through the body and capable of rendering it sensitive and lively. The neo-materialist, faithful to the spirit, if not the letter of Epicureanism, will nevertheless insist that the activity and organization of neurons and other entities actually found in our brains and bodies, or artificial entities analogous to them, is necessary and sufficient for awareness.

For the neo-materialist, the complexes we call 'organisms' have appeared on earth in the course of evolution. All are able to respond to their environments in ways that preserve their lives and reproductive capacities. Plants send out roots towards sources of moisture, bacteria follow gradients of sugar or other nutrients, and many of the larger animals engage in elaborate routines of foraging, hunting, socializing, mating, caring for offspring, and planning, following the directives and invitations of the environment.

Animals must be aware of the condition of their own bodies and the relationship of their bodies to external things, and be able to monitor changes in these conditions and relations. Consciousness is just such a presentation of the self-in-the-world, and conscious sensations, perceptions, and emotions are essential devices for steering the body around. No one has successfully explained how neuronal organization and activity gives rise to consciousness, but we know that wakeful awareness is associated with certain patterns of global excitation in the brain, dreaming with others, and dreamless sleep or anaesthetic-induced unconsciousness with still others. These and other correlations lead us strongly to suspect that an incorporeal soul is not necessary for experience, and that physical structures and processes of some sort are sufficient.

The objection is sometimes made that neuronal organization and activity are insufficient for consciousness on the grounds that we can imagine an unconscious material machine that was indistinguishable in its behaviour from a conscious human being. This 'zombie', it is imagined, could steer itself around in the world without consciousness and respond appropriately in view of its exquisite internal machinery. It might even have a brain that looks just like one of ours. This thought experiment seems to imply that consciousness is not a feature that has simply arisen in the course of evolution whose appearance is 'explained' by its usefulness, let alone its necessity, to the organism. Rather, so the argument goes, it must be an extra feature superadded to the material body with no functional significance.

The zombie thought experiment does not, however, refute the neo-materialist thesis. It shows that we can imagine brains like ours that are unconscious, not that consciousness is in fact independent of the brain. At the same time, we know that actual human beings, when unconscious, are utterly helpless in navigating in the world. The modifications that would need to be made to the human brain and body to render it capable of navigating the world in the absence of sensory awareness are unimaginable.

Why should we believe that an artificial machine lacking consciousness but exhibiting all the behaviours of a human being could ever be constructed? And how would we know that, in building the complex machine, we had not brought a conscious being, albeit one made of different materials, into existence?

Nevertheless, whether the physics required for consciousness pertains to the molecular or to the subatomic levels described by our post-Epicurean scientific ontology is still unknown, and we may be in for some surprises as the neurosciences continue to develop.

Chapter 6
Religion and superstition

Epicurus' critique of popular religion

Epicureanism was the only sect of ancient Western philosophy to deny that the gods were active in the world and influenced the course of human events. Its recovery and reworking in the context of Christian monotheism in Europe was accordingly complex. As well as further development of its critique of religion, there was pointed resistance to the Epicurean challenge.

Theists typically regard gods as the creators of heaven and earth who have fashioned it and stocked it according to plan. In archaic religions, the gods control the seasons and the weather and bring about thunderstorms and earthquakes. They direct human destiny, saving some people from shipwrecks while dooming others, favouring some and punishing the rest, and they chastise entire populations with wars, famines, and plagues (Figure 6). They determine what is right and wrong, and, the faithful believe, enforce morality by ensuring that the wicked receive their just deserts in this or another life, while the good are rewarded. The purviews and roles of the gods, their role in history, and the record of their interactions with mortals, are set out in sacred texts such as the Bible and the Qur'an. Religious believers are enjoined to revere and to serve the gods by ceremonial performances and sacrifices, the use of symbolic objects, and by deference to priests.

6. Neptune, the Roman version of Poseidon, the Greek god of the sea, storms, and earthquakes.

For the Epicureans, the gods neither create nor evaluate. Inhabiting the intercosmic spaces in a state of blessedness, and enjoying atomic but immortal bodies, they have no perceptual access to any worlds nor influence upon them. They had no role in designing our world, and they have no role in running it, or in monitoring, rewarding, or punishing human actions, nor in deciding on the world's termination. They do not move the heavenly bodies around. Insofar as the gods do not care about us or interact with us, religious observance cannot affect them in any way.

Was Epicurus, one might wonder, really an atheist who denied the existence of the gods or only a critic of the conventional religious practices and popular beliefs of his time? He refers repeatedly to the 'blessedness' of the gods, and Diogenes Laertius assures us that he was personally pious and often prayed in the temple. According to his expositor in Cicero's anti-Epicurean dialogue, *On the Nature of the Gods*, Epicurus maintained that the gods exist 'because nature herself has imprinted a conception of them on the minds of mankind'. This belief, he thought, arises spontaneously in the mind and does not require authority, custom, or law to preserve it from generation to generation.

Epicurus' orientation to religion is accordingly somewhat mysterious, and unfortunately most of our information about it comes through his critics. Did he mean that belief in the gods, though mistaken, is natural and can be causally explained? Did he mean that ideas or conceptions of the gods, rather than the gods themselves, affect people for better or worse? Was his temple-going merely a performance aimed at avoiding censure and trouble for his school, as some ancient historians thought? Alternatively, was Epicurus one of those people who today describe themselves as 'spiritual but not religious'? The key to his theology in any case is his disregard for divine power and intelligence and their application. The gods can be called 'blessed' because they are free of worry, fear, and vexation and because we should aspire to live lives like theirs as far as possible.

It is hard to see how Epicurus can make any claims whatsoever about the gods, including their location in the intercosmic spaces and their states of mind, if he believes that sense-experience is the criterion of truth. Of course, he also makes claims about the multitude of other worlds which are neither visible from here nor detectable in any other way, but the other worlds are like our own cosmos, while the gods, with their indestructible bodies and abodes in space, are unlike any other living beings. Further, all the minds endowed with reason and purpose that we know, Epicurus pointed out, have been embodied in human beings. All this makes his references to the gods seem rather tongue in cheek.

Indeed, in line with his programme of explaining everything by reference to material and environmental causes, Epicurus was interested in explaining why humans everywhere form their beliefs about these unseen entities. According to the account given in Cicero's dialogue on this topic, the Epicureans believed that the gods were not perceived by the senses but by the intellect, via images arising from the 'innumerable atoms' that compose thoughts and dreams. While some commentators appear to believe, on the basis of a problematic preposition in Cicero's text, that these images flow *from* the gods, in the manner of the ordinary *idola* emitted from solid objects, this does not seem to be what Epicurus had in mind. Rather, the texts suggest that our thoughts flow *to* the gods on account of the images.

Lucretius intensifies the critique

Lucretius' reverential invocation to Venus at the very start of his poem, and his several worshipful references to the goddess later on, belie his severe critique of religion. Venus represents a concept, not a supernatural being, and the other gods of the Roman pantheon receive no such favourable mention. Lucretius portrays the popular conception of the gods as based in dreams and imagination, and he condemns religion as a deleterious influence on human life. His forerunner, he declares, was 'the first who dared to lift mortal eyes' to challenge religion. He saw 'human life...grovelling ignominiously in the dust, crushed beneath the grinding weight of superstition'. Evidently he had access to Epicurean texts that were more censorious than those we know.

Lucretius fills in the psychological account of religious belief, maintaining that 'The minds of mortals were visited in waking life, and still more in sleep, by visions of divine figures of matchless beauty and stupendous stature'. Because these images never appeared to age, it was supposed that they were immortal. Further, people observed 'the orderly movements of the heavenly bodies and the regular return of the seasons of the year', which they could not explain, leading them to assign the gods responsibility for them, along with a dwelling place in the sky. Thunderstorms and earthquakes caused people to cower in fear of punishment for some misdeed of which they had little idea: 'O hapless humanity to have attributed such happenings to the gods, and to have ascribed cruel wrath to them as well! What sorrows did they prepare for themselves, what wounds for us, what tears for generations to come!'

In times of crisis, it is useless to appeal to the gods for deliverance. Lucretius' conviction on this point is clear in the last pages of his poem, which is concerned with the visit of the plague to Athens in 430 BCE. In his account, borrowed from the historian Thucydides, of this devastating and fast-moving contagion—probably a haemorrhagic fever similar to that caused by today's Ebola virus—Lucretius describes the piles of unburied corpses lying in the streets, the desperation of the relatives of the sick and dying, and the abandoned temples, where 'neither the worship of the gods, nor their divinity counted for much'.

The Epicurean hope is that the provision of a scientific account of the origins of the cosmos and of life, and an account of the causes of belief in powerful supernatural persons, can overcome the human tendency to believe that gods are involved in human life. Yet Lucretius realizes that this is a struggle, a tug of war between the forces of religion—with their attempt to bind people under false doctrines and superstitious sanctions—and the truth, and that it will be difficult to resist the former:

> The time may come when you yourself, terrorized by the fearsome pronouncements of the [poets and priests], will attempt to defect from us. Consider how numerous are the fantasies they can invent, capable of confounding your calculated plan of life and clouding all your fortunes with fear. And with reason; for if people realized that there was a limit set to their tribulations,

they would somehow find strength to defy irrational beliefs and the threats of the fable-mongers. As it is, they have no way, no ability, to offer resistance, because they fear that death brings punishment without end.

Rethinking God

There was sporadic approval of some Epicurean doctrines by the Fathers of the Early Church, but for the most part the encounter between the early Christians and Roman Epicureans was a formidable clash, as might have been expected.

The promise of an afterlife to the followers of Jesus and compensation for their sufferings won numerous converts before, during, and after the fall of the Roman Empire. Christianity assigned a positive value to pain and martyrdom, and St Paul, who lectured to Stoics and Epicureans in Athens in an attempt to convert them, and later St Augustine and Father Lactantius, excoriated pleasure-seeking and disregard for Providence, insisting on the sinful nature of man and the absolute power of God. The loss or restriction of literacy and literature in the Middle Ages in Europe made the clergy the preservers of philosophy and culture and, as noted, Lucretius' text largely disappeared from view, along with most of the letters of Epicurus.

An important though slender strand of 'Christian Epicureanism' appeared in some of the popular writings of the mild and urbane 16th-century Dutch philosopher Desiderius Erasmus and in Thomas More's *Utopia*. At about the same time, artists, especially in Northern Europe, turned their visual attention away from sacred and historical subjects and towards the everyday world, depicting musical instruments, flowers, foodstuffs, textiles, and other objects with sensory appeal.

The Epicurean critique of religion, combined with the Epicurean accounts of the self-formation of the cosmos and the spontaneous emergence of living forms on earth, had a significant impact on European philosophy of the 17th and 18th centuries.

As observed earlier, there was a decided attempt at this time to articulate the notion of a creator God of infinite power whose responsibility for the world is exhausted in the initial instantaneous act of creation, in which, according to Descartes, matter is created along with the laws of motion, and along with the 'eternal truths' or the laws of logic and mathematics. The patterns of the cosmos evolve according to the laws of nature. Descartes had almost nothing to say about any further actions of God, though he paid lip service to the story of Adam and Eve, and purported to be demonstrating the immortality of the soul, a challenging task set for philosophers by a Pope with definite Epicurean leanings, Leo X.

Baruch Spinoza adopted an even more radical stance, arguing that God was nothing over and above the corporeal universe, as the mind was nothing over and above the body. As a 'finite mode', the individual was worn down by the forces of nature and subject to destruction. Despite the suggestion of Epicurean influence on these topics, Spinoza does not qualify as an Epicurean, as he denies that particles of matter are fully real and indestructible—only the universe as a whole is eternal. Further, his determinism and his conception of the emotions as undesirable disturbances link him to the Stoic tradition.

In other early modern metaphysical systems, the Genesis account of Creation was set aside in favour of imaginative non-Epicurean variations. For Leibniz, an infinite universe of soul-like entities emerges as a viable combinatorial package from the mind of God. For Berkeley, it is an array of minds with volitions and ideas that are created. With the exception of Thomas Hobbes, about whom more in Chapter 8, and Spinoza, all the major metaphysicians of the 17th century retain God as the author of moral truths as well as physical reality. At the same time, they are disposed to accept the Epicurean premise that God does not interfere in the world, declaring that He does not suspend the laws of nature to produce miracles, or at least has not done so since Biblical times, and that all that happens according to the laws of mechanics has been foreseen and approved from the beginning.

Immanuel Kant followed suit in this respect. In his early work, the *Universal Natural History* of 1755, admitting to his readers the close correspondences between his views and those of Epicurus and Lucretius, Kant presented a mechanistic account of the formation of the cosmos and the living inhabitants on our planet and on all the other planets. God was said to be the author of the few and simple laws of nature that sufficed to produce the universe. It ran like clockwork, with entire planetary systems coming into existence and passing away in a very Epicurean fashion.

With the basis for the Christian revelation in miracles no longer secure, and with doubts multiplying as to the authority of Scripture, the Epicurean alternative, implying a final and decisive mortality and an ethics arising out of the requisites of social life, became increasingly available.

These thought pathways were developed further in 18th-century philosophy. In the French Enlightenment, an Epicurean atheism attached to a basically Cartesian faith in the inexorability of the laws of nature reappeared in the writings of Paul-Henri Thiry, Baron d'Holbach. Holbach repeated all the elements of the Epicurean religion-critique: theology, he said, is but 'ignorance of natural causes reduced to a system'. Everything comes from matter; religion deals with phantoms and God is imaginary. Religion promotes cruelty and division, setting people to 'hating and tormenting each other for unintelligible opinions'.

Despite this array of alternatives to the account of creation in six days, the age-old argument against the Epicureans—that the visible world could not be a product of the 'fortuitous concourse of atoms'—was difficult to set aside. Other writers mounted considerable resistance to religion-critique in the form of the 'argument from design' especially favoured by British and Dutch authors. How could the astonishing adaptation of animals to their food sources and environment be explained except by an act of divine creation? How could their instincts be fine-tuned to their survival and propagation? Why did the species breed true to type if there were not a divine template in the mind of God that underwrote their creation?

The Epicureans had floated the notion of the persistence of viable forms and the elimination of nonviable forms, as discussed in Chapter 4. There was a growing awareness in the 18th century that new animal types had come into existence and old types had become extinct. There was also a keen awareness of how selective breeding by agriculturalists could alter flowers and livestock. But the kind of evidence and argument that Darwin would cite in his treatise *On the Origin of Species* of 1859 for variation of individual traits and characteristics in every generation, and the slow 'selection' of them by nature on the analogy of the breeder's choices, was not yet conceptually available. It took a bold intellect to argue, as Holbach did, that when it came to the beauty, intricacy, and reproductive powers of butterflies, insects, and polyps, '[a]ll these things will not prove the existence of this god; they will only prove that you have not the ideas which you should have of causes and effects that can produce the infinitely diversified combinations, of which the universe is the assemblage'.

David Hume in his *Dialogues Concerning Natural Religion*, published posthumously in 1779, attempted the difficult task of attacking the argument from design. As he pointed out, when we come across a watch, we rightly suppose it to be the work of an intelligent, skilled, and well-meaning artisan. But this is because we have experience both of watchmakers and watches; we know watches to be products of human industry that do not come into existence from nothing or from bits of metal left lying about according to the operations of the laws of nature. The relationship of a watchmaker to a watch is, however, a poor model for thinking about the origins of the world. The universe has come into existence only once, and we have never observed the construction of a universe by a deity.

The existence of evil in the world, and the apparent punishment of the innocent in natural disasters and personal accidents, was long seen as a motive to deny Providence and to adopt the Epicurean philosophy. According to Lactantius, writing in the 3rd century CE, Epicurus had asked whether god is unable to prevent evil but willing to do so, or able but unwilling to do so. Either alternative would seem to condemn the divinity. Apologists for religion, notably Leibniz, took up the problem, arguing that many evils were only apparent and not really evil, or that they were necessary for the greater long-term good. Hume pursued the question relentlessly in the *Dialogues*, painting a bleak picture of the world as disorganized and tragic. If one is going to have recourse to supernatural entities, he argued, one might as well suppose the world to have been created by a semi-competent deity, or to be a rough and badly flawed first attempt at world creation.

Hume was attracted to the materialism of his day, but his sceptical orientation as well as some concern for his literary reputation and professional prospects prevented him from advocating for it openly. He succeeded nevertheless in undermining religious faith while removing all questions of ultimate origins from the realm of decidability. Kant, who reported himself awoken by Hume from his 'dogmatic slumber', was shocked by the *Dialogues*, though he was as ready as Hume to dispense with the mythological features of Christianity. After his so-called 'critical turn', probably worried about the moral implications as well as the epistemological status of his grand vision of *cosmoi* mechanically emerging from chaos and passing into it again, Kant argued in favour of a ban on theorizing about the ultimate origins, constitution, and fate of the world on the grounds that such questions were outside the scope of human knowledge. The materialist account of the origins of the world was just as philosophically indemonstrable as divine creation, he decided, so one ought to affirm neither account.

In Kant's view, Hume's attack on religion was insensitive to human needs; Hume had not hesitated when it came to depriving people of comforting thoughts concerning Providence and the rewards of virtue. So 'God' became for Kant the name, not of the author of the moral law, but only of an idea that he thought people ought to hold before their minds to provide a kind of moral orientation. God was the consoling idea of a power that could and would reward virtue whether the world did so or not.

What's wrong with religion, one might wonder in taking up Kant's suggestion, if it motivates people to act morally, even if they will not in fact be rewarded with immortality for doing so?

Lucretius' answer to this question had been that religion prompts cruel actions. People will persecute, murder, and mutilate others whom they regard as ungodly, or because they are under the impression that God wishes them to do so, and these brutal individual impulses will be accepted and reinforced by one's inspired co-religionists. Near the beginning of Book I of his poem, Lucretius describes in poignant terms Agamemnon's sacrifice of his daughter Iphigenia, 'a sorrowful and sinless victim of a sinful crime'. Her father's aim was to placate the goddess Artemis in order to change the winds that were preventing Agamemnon's fleet from setting sail for Troy. The episode reveals both Lucretius' anti-war sentiments and his detestation of 'religio'. 'Such heinous acts could superstition prompt'.

Is religion obsolete?

The mysteries associated with diseases, natural disasters, and sudden deaths have been explicated by modern science in ways the Epicureans would have approved of, and geologists, botanists, biologists, and geneticists have given promising if still speculative accounts of the origins of life and the mechanisms of speciation. Cosmology has several theories at its disposal for explaining the origins of the universe, including the Big Bang. One might wonder whether Lucretius was correct in believing that a science of nature could never decisively eliminate religious belief.

To appeal to the power of priests, as Lucretius does, to explain the pull of religion only pushes the question back a step. Why would the priests have such authority if their utterances could be seen to reflect scientific ignorance and if persecution in the name of religion was really foreign to human nature? To the extent that one has been exposed to or practices a natural science, one is less likely to accept, at least literally, the fundamental teachings of religion, or to believe that people go on to exist in another world after they die, where they are punished or rewarded. Yet many people who are sophisticated in geology, physics, astronomy, physiology, biology, and so on prefer to describe themselves as 'agnostic' or 'spiritual' rather than 'atheistic'.

There are surely better explanations for this preference than the terrifying utterances of priests. Materialists and others who describe themselves as spiritual but not religious may want to express their admiration for the productions and ceremonies of the past—the glorious paintings of Biblical themes, the architecture and sculpture of temples and cathedrals, the emotional power of sacred music. They may feel that to discard religion is to demean these aesthetic productions and to be unable to enjoy them fully. Religious services offer many people an opportunity to consult their consciences, and to enjoy the experience of being in a peaceable group with a shared focus. The worshippers are together, and they are all on the same side, not angry, as in a riot, nor separated into rival team supporters, as at a sports match, and they are not engaged in chatter, as in a quilting circle. Religion presents a world apart from the everyday one of moneymaking, consumption, and family responsibilities. Its ethical requirements are usually clearly stated and easy to understand and remember. Ancient Greek and Roman priests were not involved in such morally worthy activities as the education of prisoners, the care of the sick and the orphaned, and resistance to oppression and war, activities that have to be chalked up to the credit of at least some later religious institutions.

As the Epicureans understood, there are material causes as well as cultural reasons for participating in religious life. Religious experiences, beliefs, and practices are phenomena in human life that can be explained. Neuroscientists believe they have found areas of the brain that are particularly prone to religious ideation, as was suspected at the end of the 19th century by the psychologist William James. Where the Epicureans believed that images of the gods were assembled by the dreaming mind from drifting streams of *idola*, we now strongly suspect that it is the mind itself that generates the feelings of being beholden to a god and participation in a divinely run universe. These feelings likely play a role in the capacity for moral restraint upon which the biological success of our species depends.

Epicurean philosophy offers nonbelievers a general theory of the natural origins of life and the universe, a respectful account of morality as partly a natural disposition, partly a 'social technology' improved by experience and reflection, and a chance to dissociate from religious institutions whose history and practice of unkindness and persecution they may deplore. Alternatively, the non-believer may remain within religion, as unmoved by threats of divine punishment as by promises of divine reward and fully reconciled to the finality of death, but moved, as Epicurus himself appears to have been, by ideals of piety and community.

Chapter 7
Politics and society

From status to contract

Justice was a revered condition and concept in ancient political philosophy. The term had a broader meaning than we assign to it today, referring to wrongdoing in general. Not only institutions, proclamations, and legal and governmental decisions and pronouncements could be described as 'just' and 'unjust', but also the characters and actions of individual persons.

Justice, said Epicurus, has a single basic function: to make people useful to one another and to prevent one person from harming or being harmed by another. It is a contract, a non-aggression and cooperation pact, instituted between human beings who are dependent on one another but also dangerous to one another. There is no justice due to or attainable from animals, he claimed, because they cannot enter into the formation of such contracts with humans. Any human being who is so irrational as not to be able to make and keep such compacts, should accordingly be treated as a dangerous animal.

Although the conception of justice as a system of harm reduction that protects the weak from the strong was not unknown to ancient philosophers—it is the target of Callicles in Plato's dialogue *Gorgias*—it was not defended in the other main schools of philosophy. Justice in ancient political philosophy was more commonly understood as a social condition in which each person behaves appropriately according to his or her role in the wider society, and each is given his 'due'. It was taken for granted that people belonged to different categories of human being and were owed different things by others. Citizen and non-citizen, master and slave, husband and wife, each had different rights (or none) and responsibilities. In his description of an ideal city-state in his *Republic*, Plato distinguished between the gold, silver, and bronze people, emphasizing a clearly defined division of labour and different levels of human worth within the community.

The ancients understood the contingency and reversibility of some social roles, including that of master and slave, or citizen and foreigner. Slaves were sometimes freed on the deaths of their masters, and some might even become quite wealthy. Aristotle realized that a king could become a slave if taken in a war, a situation he described as 'unjust'. But outside of the exceptional cases that proved the rule, slavery was seen as just, and indeed by Aristotle as grounded in nature. The social dominance of men over women was considered equally natural. Under later Roman law, masters had the right of life or death over their slaves, as did fathers over their children. Roman fathers could decide whether their children were to be raised or abandoned, and their mothers were legally powerless in this regard.

Plato's only egalitarian initiative had been to suggest, to the consternation of some of his readers, that males and females might receive similar educations and training, and that women might be admitted to the top class of Guardians. Many thought he was joking. Aristotle, who was more influential than Plato on this topic, understood difference and domination as fundamental to the structure of the universe and important to reinforce socially. Males corresponded to 'form' (one fundamental ontological category), females to 'matter' (the other). As form governed matter, through its intrinsic superiority, and as certain divinities, the Intelligences, governed the celestial bodies and so the weather, the seasons, and the life cycles of plants and animals, males governed females and masters governed slaves.

All had their natural roles and places, and functioned in an optimal social order. There was something in the essential nature of the slave (his or her inability to direct action and his or her exceptional biddability), and something in the nature of women (their lack of rationality and their emotionality), which made them unfit for independent life. Women were necessary for the perpetuation of family lineages and responsible for clothing the family, but the civic order was believed to be best preserved if citizens' wives remained for the most part indoors. The qualities of a good woman (chastity, a docile temper, and industriousness) were different from those of a good man (courage, a sense of personal honour, and correct public demeanour).

For the ancients, then, people were not equal in the eyes of the law; they did not occupy interchangeable roles and were not possessed of general 'human rights'. For the most part, philosophers were concerned to explain the utility and necessity of these social categories, which they regarded as originating from nature.

The Epicureans were exceptional in this respect. Epicurean theory has no place for natural hierarchies, only degrees of complexity. Insofar as everyone is made of the same atomic material and nothing else, all social relations depend on human perception and convention. The domination of one person or many by another, or of the many by a few, is only a social fact in Epicurean ontology. The notion of giving people occupying different social roles their 'due', or what they intrinsically deserve, can therefore have no metaphysical basis.

At the same time, ancient materialism provided a basis of its own for demeaning comparisons. Temperament and capability must depend on bodily structure; just as foxes and rabbits have different dispositions and habits, so do men and women. So Lucretius, for all his interest in love, generation, and renewal, could be frankly disparaging about the physical and mental powers of women, insisting that, with individual exceptions 'the male sex is, generally speaking, far superior in skill and ingenuity', at least when it comes to the invention of the tools of civilization (including the loom). This view, needless to say, is not held by anthropologists, who have given rich accounts of women's artistry and technological innovation in early times and in the few remaining subsistence cultures.

As sense perception is the ultimate test of truth for the Epicurean, preference and choice are the ultimate test of moral and political goodness. Preferences and choices differ from place to place and can change. If people's awareness and estimation of their needs and the possibilities open to them changes, their laws and customs must change as well; there is no absolute justice that is right for all times and places. Further, all societies evolve. On the atomist scheme, the social order of any given era is temporary and has arisen through a combination of chance, necessity, and human free will, all of which interact to produce the current state of affairs. As no configuration is permanent, and the perpetual, restless movements of the atoms is unceasing, we can expect change and even revolution. No city, empire, or social system can be expected to last forever. Most importantly, we must never suppose that the society in which we live has been intentionally designed by a wise author, or that it represents an optimal adaptation to the circumstances. We can only suppose that, as with animal forms, some arrangements will collapse from their own internal instability or their incompatibility with their environments.

The 'selfish' nature of the Epicureans was noted by their critics. However, this selfishness did not relate to their egoism as individuals, but rather their avoidance of political engagement and responsibility as presenting too much of an opportunity for vexation. For the Epicureans, politics implied a striving for power and admiration which was incompatible with virtue, pleasure, and peace of mind. These goods were only to be found in private relationships with like-minded individuals. The location of the Epicurean garden outside the city boundary symbolized this reclusive ideal, and the Epicureans distinguished sharply between coexistence with the strangers and mere acquaintances of one's city, and true friendship, in which relations are governed by affection and trust.

In this regard, they departed from the Stoics, who thought that love of friends and family could be extended into civic friendship and kindly regard for one's neighbours and fellow citizens, and ultimately flow towards distant strangers in an 'expanding circle'. The contrast between Stoic and Epicurean moral philosophy will be treated more fully in Chapter 9.

In light of their preference for living apart, one might suppose the Epicurean influence on later political theory to be weak or non-existent, but this is not the case. There is debate as to whether, in its own time, Roman Epicureanism was a populist movement, implying a scepticism about authority that was regarded as dangerous by the patrician class. As Epicureanism was rediscovered in the early modern period, conflicts over conceptions of political authority and justice as they had developed in Christian-Aristotelian culture became visible. Epicurean themes began to emerge through the old fabric, profoundly altering political thought and arrangements.

The evolution of civilization

After the fall of the Roman Empire in the 5th century CE, European societies consolidated into systems of monarchs and nobles, serfs, artisans, the clergy, the military, and traders. Political institutions were described in terms of 'natural law' and the 'divine right' of kings to tax, punish, and wage war.

Where the deprivations of the poor were concerned, the view was propounded that poverty and famine were the will of God and that the deprived would be compensated in heaven for their sufferings on earth—or would even, according to the Bible, inherit it.

Misery was the lot of the multitude and even of many of the better off, on this interpretation of Scripture, because life was a test of worthiness and punishment for sin. Adam and Eve had been the divinely created sole inhabitants of the earth before sin entered the world. The pair had lived blissfully until they disobeyed their Maker, and were expelled from the Garden of Eden and forced to toil. Although the Bible contains no explanation of how early societies organized themselves into rival monarchies and is ambivalent about kings and kingship, Christian writers found it necessary to acknowledge the temporal order. With St Augustine and his successors, kings, magistrates, and priests with the power to command and direct human society were conceptualized as divinely appointed and hence due obedience.

The Epicurean account of human history was altogether different (Figure 7). Humans, recall, were sprung from the earth like other large creatures, in the period when the earth was still fertile and able to produce wombs for life. The earliest humans, Lucretius proposed, were wild, shaggy creatures living solitary lives in caves and forests. As time went on, they found themselves threatened by wild beasts and other dangers and began to band together with others. Language and clothing were invented, and people began to form family groups and later tribes and kingdoms. Rather than being taught the arts such as weaving, music, and writing by the gods, as mythology posited, they learned from observing the other animals and employing their own inventive powers. Thus they raised themselves by degrees to a condition of civilization.

The ascent was at the same time a bloody and terrible one. Observing how forest fires caused molten ores to run out of the

7. Early humanity on the verge of discovering the use of fire and so agriculture and warfare. Detail of Forest Fire, c.1505, by Piero di Cosimo.

ground, people discovered the beauty of gold and silver and hit on the idea that they could work the harder metals into any form they chose. So they fashioned axes for felling trees, ploughs and sickles for agriculture, and also bronze and iron weapons, far more lethal than the stones, clubs, and bare hands of their ancestors. Warfare appeared on earth and with it monarchy, political rivalry, and assassination. Life was insecure and trust was impossible. At length, Lucretius argued, humans grew tired of constant strife and imposed laws and enforcers. Had they not done so, the human race would not have survived. What we call 'justice' is therefore a human invention that did not exist until human beings decided to propound laws and try to ensure obedience through authorities invested with punitive power.

For the materialist Thomas Hobbes, working and writing in England in the mid-17th century, the Epicurean account of the invention of justice was a prompt to his own theory of the social contract. Hobbes described an original condition of 'war of all against all' that could be terminated only by the renunciation of the 'natural' right to appropriate anything one could and to pursue one's self-interest without restraint. Although Hobbes agreed that the basis of justice is human agreement, and that its function is to prevent harm to others for one's own advantage, he did not think individuals could abide by their anti-aggression covenants with other individuals in the absence of a central power to punish them if they lapsed. He added to the Epicurean account an enforcer, his 'Leviathan', or absolute ruler, in his treatise of that name published in 1651.

It was evident from Hobbes's text, however, that the role of this ruler was simply to enhance the prosperity and comfort of the citizenry and to protect them from threats and dangers; in other words, to remove fear and pain and increase pleasure. The sovereign's role was not to cultivate personal glory, or to extend his or her territory via conquest, or to make court life splendid. In this regard, as well as in his evocation of an anarchic state of nature, Hobbes showed his Epicurean colours. He recharacterized the 'laws of nature' as being pragmatic rules for reducing social friction given natural human psychological tendencies, rather than the moral and political commands of God. Other major political theorists of the 17th century, including Hobbes's forerunner Hugo Grotius and later Samuel von Pufendorf, appealed to 'natural law' in this non-theological sense.

The social contract theory of the French philosopher Jean-Jacques Rousseau, developed a century later than Hobbes's, denies the need for an absolute monarch. Rousseau assumed, in Epicurean fashion, an absence of status distinctions between the contracting parties and a mutual interest in peace. The law should depend on what those affected by it want and can agree to; it is contrived from the bottom up by reference to the desires and preferences of the people rather than decreed from the top down according to the preferences of the ruler.

These adaptations of Epicurean political theory have profoundly affected modern institutions. The evolution of law in European societies was persuasively argued by Henry Maine in the late 19th century to be a progression from relations based on status to relations based on contract. In the latter paradigm, persons are treated as interchangeable under the law. No special privileges, for example, in the form of immunity from prosecution for ordinary crimes are accorded the wealthy, the well born, the propertied, or the clergy. People are free in principle to leave their place of employment, to change their husbands as well as their wives, to take up any profession for which they have received the appropriate qualifications without regard to religion or ethnicity, and to perform other free actions. The notion that legislation is meant exclusively for the general welfare is widely accepted. Today it is difficult to pass laws that benefit only the few without persuading the many that the measure is intended to benefit them. Unfortunately, it is often all too easy to persuade them accordingly.

Progress and its problems

The Epicureans recognized, as their philosophical rivals rarely did, that the progress of technology and social organization that had occurred over many centuries had imposed not only gains but also losses. All the refinements of civilized life—here Lucretius mentions roads, buildings, fabrics, paintings, and sculpture—contribute to the enjoyment of life. At the same time, under civilization, men's minds are poisoned with ambition and greed, evils that arise from 'not knowing the limits of possession'. As jealousy vitiates relations between intimates, political and financial aspirations produce bloodshed and civic strife.

In his celebrated *Discourse on the Origins of Inequality*, published about twenty years before the French Revolution, Rousseau followed Lucretius' account of the moral deterioration that accompanied the technological progress of the human race. As

Lucretius deplored the growing taste for luxury in people's disdain for acorns, grass beds, and fur and leather garments, observing that 'gold and purple...plague human lives with cares and weary them with war', Rousseau attacked the commercial state whose superfluities and delicacies produced slavery and misery. Civilized man seems to be 'always moving, sweating, toiling, and racking his brains to find still more laborious occupations'. Lucretius was dismayed by warfare, which developed as an institution only with metal technology; Rousseau found 'more violent outrages in the sack of a single town than were committed in the state of nature during whole ages on the whole earth'.

The imagination of writers of the period was stimulated by the concordance of Lucretius' descriptions of prehistory with the discovery of 'savages' in North America, Tahiti, Australia, and New Zealand. There could be little doubt now that the Genesis account was highly unlikely and the Epicurean account substantially correct. Opinions varied as to whether the lives of these found people were, as Rousseau surmised of our hunter-gatherer ancestors, happier and less vexed and exploited than those of his contemporaries, or whether, as proponents of progress like Adam Smith and the Marquis de Condorcet proposed, the world was moving from a primitive, uncultivated state in which force ruled, towards a condition of universal opulence, equality, and enlightenment.

The Epicurean notion that human societies had passed through a series of distinct economic and social phases, driven by technological change, was further developed by Friedrich Engels and Karl Marx, writing in the mid-19th century for German and English readers. Marx, who had written a doctoral dissertation on the epistemology of Epicurus in relation to his atomist predecessor Democritus, felt no Lucretian-Rousseauist nostalgia for a lost golden age of pastoral life. The communists looked forward to a new era, ushered in by the technologies of mass production and the factory system.

The worker, Marx and Engels believed, had been exploited and degraded to the condition of an animal in the pursuit of profit and luxury by the capitalist class. In time, however, all would be reconfigured by natural mechanisms. The capitalist system, like a species whose time is up because its parts can no longer function to sustain it, would become extinct, giving way to another, more sustainable mode of the organization of labour and the distribution of profits. The working day would be shortened and the worker would be freed for leisure—and pleasure.

Marx's predicted further stage of human history has alas not yet appeared. Instead, technological innovation has fed the appetite for more consumption and more labour, not for leisure or a return to simple, costless pleasures. The 'limits of possession' have not been taken to heart. Recognizing this, the anthropologist Marshall Sahlins and other contemporary writers have pointed to the ways in which the lives of the few remaining hunter-gatherers are healthier and happier than those of the citizens of modern industrial societies. It is not only survivalists, proponents of the Palaeolithic diet, polyamorists, and other minor sects who find the descriptions of the golden age by the ancient poets inspiring. The Epicurean accounting of the gains and losses of civilization in the engaging form in which Lucretius presented it in Book V of *On the Nature of Things* continues to provide material for reflection and for the philosophical imagination.

Chapter 8
Epicurean ethics

Pleasure and pain

The Epicurean moral tenets concern living, loving, and
dying: how to pass one's days and evenings, how to bear suffering,
how to manage one's love life, and how to face the prospect
of death. Their recommendations reflect the conviction that
although pain and pleasure can be felt as either 'psychological'
or 'physical', the mind is inseparable from the body, and 'all
good and bad consists in sense-experience'. The material nature
of the body and mind makes suffering and death inevitable,
and the latter final and incontrovertible. As Lucretius observes,
both mind and body succumb to the 'stress and strain of age'.
But the material nature of mind and body also makes pain and
suffering remediable, and opens us to many sources of joy
and happiness.

It bears remarking that this 'eudaemonistic' stance, focused on
personal happiness, is typical of ancient ethics, while contemporary
ethics, under the influence of Bentham and Kant, takes as its
point of departure the welfare of the other, or others, rather than
that of the self. Avoiding pain, guilt, anxiety, and fear as far as
possible is the key to the happy life for the Epicureans, and
explaining how to avoid or at least minimize them is the goal
of philosophy.

The world abounds—as we still recognize—with threats to bodily comfort, psychological ease, and indeed to life itself. In earlier times, humans faced physical threats from wild animals, and they continue to face storms and earthquakes, and diseases carried by miasmas, as well as the threats posed by civilization—industrial poisoning, traffic accidents, and warfare. The mind is subject to epilepsy, dementia, and coma, which are forms of poisoning or physical disruption. Ancient and early modern philosophers recognized that the social world is not always a fountain of goods and benefits but abounds in threats to psychological well-being. Those cited by the ancients included loss of lands, the threat of being disgraced, exiled, executed by a jealous tyrant or a corrupt magistrate, or seeing one's children, siblings, or friends die.

Epicurus observes that animals instinctively seek to avoid and escape pain and deprivation and to prolong their lives. They also seek to gratify their appetites and enjoy themselves. There are reasons in particular cases to struggle against these natural tendencies, to suffer pain temporarily and voluntarily by undergoing surgery or participating in rough sports. Further, our physiology is such that some kinds of pain—that induced, for example, by the consumption of chilli peppers, by being tattooed or incised, or by extreme sports—are followed or attended by pleasure because of the release of natural opiates in the brain.

Nevertheless, the happy life requires for the most part the avoidance or mitigation of pain. Merely escaping the 'cry of the flesh', the pain of being cold, hungry, and thirsty is itself pleasurable. It seems obvious that once basic needs have been satisfied, one should try to arrange one's life so as to be free of annoyances—whether these take the form of noise, or unpleasant surroundings, or dull company, or tiresome work—and free from frustration and overhanging threats.

Self-denial has no ethical importance for the Epicurean except as a means of preventing pain. It is nevertheless important in that respect. The direct pursuit of pleasure too easily leads to dissipation and to long-term ill effects. Epicurus is clear on the point: 'It is not drinking bouts and continuous partying and enjoying boys and women, or consuming fish and the other dainties of an extravagant table, which produce the pleasant life, but sober calculation.' While eating and drinking are pleasurable, once the boundary between satisfying hunger and indulging in gluttony, or between the enjoyment of modest intoxication and gross drunkenness is traversed, pain and trouble are the usual results, as anyone who 'searches out the reasons for every choice and avoidance' can establish. So pleasure ought not to be pursued if the prospect of pain is in the offing, and pains ought to be embraced if they promise a relief of future suffering and greater pleasure in the future (Figure 8).

In assessing Epicurean moral philosophy, one might think there is little to be said against its prudent view of minimizing pain while enjoying moderate and sensible pleasures. But prudence seems

8. **Epicurean sensuality frowned upon by many moralists found ample expression in European art.**

different from morality or virtue, and one might feel that Epicureanism is too egoistic and calculative in focussing attention on the future consequences of actions for the agent rather than on the intrinsic rightness of acts and their effects on other people. What about the virtues of honesty, generosity, and respect for the rights of others and their property?

The Epicureans noted that we have to live with others, including others whom we do not especially like, and towards whom we have no warm, generous, or tolerant feelings. We should behave justly towards them, as well as prudently towards ourselves. But what makes the social virtue of honesty 'morally good' is not a mysterious intrinsic property of truth-telling. It is that others expect the truth from us and resent our lies. Experience generally shows that we can expect to be shunned and punished for dishonesty. On occasion, an individual may get away with an action of the sort that would be punished if detected. But no one can reasonably expect to go through life telling lies, thieving, hoarding, and insulting and injuring others with impunity. Such a person must always fear disclosure and retaliation and have accordingly an unpleasant life. In this way the Epicureans were able to defend the traditional virtues as worthy of pursuit, while still maintaining that all moral motivation was ultimately based on the avoidance of pain.

Desire and disappointment

The most painful and pleasurable aspects of human life are those related to love and passion. The Epicureans were reported to have written a great deal about sexual desire, in a fashion that rival philosophers considered unseemly. Epicurus had many female friends, including what are now termed 'friends with benefits'. He warns against love, and his attitude to sex is best described as recreational. Sex is natural, he maintains, but it is not necessary. He advises his pupils to choose their partners according only to age, beauty, and figure, rather than, as convention dictated, wealth and birth, and he warns his pupils that children can be a lot of trouble. The wise man, according to Epicurus (addressing his male readers), generally avoids marriage and procreation, but for some it may be suitable.

Lucretius has a somewhat different orientation. Marriage, he thinks (again addressing his male readers), can succeed with a woman who is neat and obliging. Familiarity breeds love as 'drops of water falling upon a rock in long lapse of time' hollow it out. His defence of marriage as a reasonably pleasant condition, as mild as it is, is noteworthy. At the same time, the metaphors of limits and renewals are woven through his poetry, and make a spectacular appearance in his moral psychology.

There were particular reasons for Epicurus' dim view of romance and Lucretius' warnings against it. The Athenians of Epicurus' time differentiated the class of educated, companionable, courtable women sharply from both ordinary prostitutes and wives acquired for strategic and procreative reasons. Men of means competed for the attention of desirable courtesans with gifts and money. Men are changeable and do not have bottomless purses, and for the sake of her own security a courtesan had to keep her eyes open for her next or additional conquest. All this produced a good deal of anxiety, resentment, and perhaps disgust on the part of her admirers, and understanding this feature of the social system helps to put the Epicurean recommendations to men to beware of emotional attachments to women into perspective.

For Lucretius, although love for individuals and the desire it engenders is hounded by futility, and although the bonds between individuals are, like everything else, eroded by time, love is indispensable in the cosmic scheme as a force of reconstruction and renovation. Mortals are 'beckoned by divine pleasure' to replenish the world with their kind, and women are as active, involved, and important as men in this regard.

In Book IV, he describes lovers as insatiably gazing on one another and covering one another's bodies with caresses, trying unsuccessfully to merge their bodies to correspond to the unity they feel. But passion also undoes its human victims. Love is the 'honeyed drop of Venus' sweetness that is first distilled into our heart to be followed by chilling care'. Obsessive thinking about the beloved, and anxiety and anguish in the form of the fear of loss or loss itself stalk the lover. Reputation and business suffer, funds are depleted by present giving. Apart from the desperation of madness, with which it has much in common, the most painful emotion experienced by humans is sexual jealousy, which arises from the impression that we must have the affections of one person in particular and that no one else can take their place. The 'ambiguous word' thrown out by a lover or the 'trace of a smile' directed at another can propel us into a state of anguish.

Lucretius turns somewhat satirical on the subject of these passions. They can fasten on the short, the overgrown, the plain, and the bad tempered, who appear to us dainty, imposing, lustrous, sweet, and brilliant. The remedies for jealousy and lack of possession are to think on the defects of the beloved and their bodily nature and to remember that 'we have lived without her until now'.

As Epicurus recommended, avoid contact with the person causing you pain (regardless of the short-term pleasure you derive from the association), and go out of your way not to see them or hear their name mentioned. Further, when in the grip of a painful romantic obsession, Lucretius recommends sex with 'random-roving' Venuses you don't love for relief and diversion. Much of what Lucretius has to say on this score is insightful and accompanied with good advice, even for systems of mating organized differently from the Athenian or the Roman.

It is not only the appetites for food, intoxicants, and sex that need to be moderated for the Epicurean, but the lust for wealth, fame, and power. Normal human beings require some money, reputation, and control over their environment to be happy, but their appetite for these things may overreach what sober calculation would recommend. The lives of the very wealthy, famous, and powerful are, as we know from the tabloids, subject to dramatic reversals of fortune, by way of lawsuits, scandals, painful dissolutions of marriage, addictions, and suicide.

In some respects, Epicureans and their rivals among Platonists, Aristotelians, and Stoics, were not as far apart as might be imagined. Freedom from disturbance, the achievement of a form of tranquillity, and the avoidance of profligacy are common to all. But as important as what the Epicureans said in defending the aims of experienced pleasure and avoidance of pain was what they did not say.

Though Epicurus advised sober reflection and attention to the grounds for 'choice and avoidance', he did not demarcate sharply between natural animal behaviour and human behaviour dictated by 'reason'. The Epicureans did not suppose that the virtues were fixed, unchanging action-templates existing independently of human motivations and preferences. Epicurus did not cite examples of the virtuous person acting against their own self interest except in one case: he thought it good to sacrifice oneself, even to the point of enduring torture, for a friend.

The finality of death

The Epicurean teaching on the finality of death is perhaps one of the most difficult for many human beings to accept. Unlike the adherents of the major philosophical schools of antiquity and of most world religions, Epicurus and Lucretius were convinced mortalists.

Mountains crumble, seas evaporate, all buildings collapse with time, and all plants and animals die. The atoms that once composed the body, including the material soul atoms, disperse and mingle with the atoms of earth, sea, and sky. No individual can continue to live or be revivified after the cessation of its vital functions and the natural decomposition of its body. Christian teaching demanded the resurrection of the person and his or her body, and Epicureanism was on this account damnable doctrine in the eyes of the Church.

The human mind is clearly predisposed to a belief in immortality or the transmigration of souls into other people or animals. Why this should be such a universal belief is easy to understand. The strong bonds human beings form with their friends and relatives depend on the distinct impression that another person's 'personality' makes on us. A corpse is an item from which personality and personhood has seemingly fled, and it is natural to think that it has gone 'somewhere' or is waiting in the wings to inhabit a more functional body. The behaviour of a bird, an animal, or a child can make us think that a familiar soul has settled in and is inhabiting that body.

The Christian doctrine of an incorporeal, separable, immortal soul subject to reward and punishment has been advanced by two main routes: first by argument and second by revelation. Platonic philosophy asserted that the soul was incorporeal and so indivisible and indestructible, and this view was adopted by the North African Augustine (later St Augustine) who had come under the influence of Neoplatonists when studying in Milan. It fitted with the Christian teaching that Jesus had died on the cross to expiate the sins of mankind and restore eternal life, lost in the Fall of Man. The reliability of this promise was dependent on belief in the miracles performed by Christ and his disappearance from his tomb and appearance later to his disciples.

The Epicurean theory of a fragile, material soul accordingly presented a challenge to the philosophers of the 17th century. The need and often desire to accommodate philosophy to theology produced what might be termed the golden age of metaphysics, a creative fusion of scientific naturalism with theological supernaturalism.

The Platonic notion that the soul is a separable, incorporeal substance that departs from the corpse and maintains itself intact without a body, either forever or until a new body is awarded to it, was revived by Descartes in his *Meditations* of 1640. The capabilities of this indestructible soul did not, however, include perception or memories of its past life, which he agreed depended on having a body outfitted with a brain.

John Locke, tempted by the materialists' idea that not only memory and perception but indeed all mentality was produced by the brain, suggested that an individual's body and brain, with all its reward and punishment-relevant memories, could be reconstituted at the Resurrection.

Gottfried Wilhelm Leibniz, by contrast, seeing the difficulties with both hypotheses, proposed that living individuals in fact never actually die. Nor is what we call 'birth' their beginning. Rather, before birth (indeed before conception, according to Leibniz) every human being has a miniscule living body, and after what we call 'death' the person shrinks back into a smaller format, awaiting regrowth to full size.

For the Epicureans there is no such persistence and no return. The hell of the pagans was, like the Christian hell, a dark, underground, and unpleasant place, though not as vividly imagined with demons and tortures. There is no such place, Lucretius insists, and so nothing to fear on that account, though there is no heaven to hope for either.

Having embraced mortalism, the Epicureans tried to face up to its psychological implications, bravely remarking on the unavoidability of grief and mourning in human and even in animal life, while attempting to alleviate the fear of death. But what about the often painful process of dying and the very fact of being dead? How can one fail to be dismayed by and fearful of those conditions?

As the contemporary moral philosopher Thomas Nagel has pointed out, what makes death frightening is the knowledge that we are going to be deprived of the goods of life and there is nothing we can do about it. Merely imagining that one is scheduled to die, even painlessly, tomorrow at noon, is a terrible thought. One's involuntary death would seem to be the worst possible thing that can happen to one, impossible to contemplate with equanimity. If I do not live beyond noon tomorrow I shall be deprived of so much—from my first delicious cup of coffee every morning, to the enjoyment of sunsets, movies, and new clothes, to watching my children grow up and contributing what I can to new knowledge, and so on. Regardless of the grief and practical problems my death will cause for others, death seems to be bad for me. What could be more horrible than to lose all my memories, to be unable to awaken from sleep, unable to move and speak?

To this the Epicurean argues that the deprivations cited would be horrible for a living person but that they are not for a dead one. 'When we exist, death is not yet present, and when death is present, then we do not exist'. We imagine ourselves lying paralysed, blind, and deaf in a dark tomb and feel horror and pity. But we will not be the observers of our deaths, and there will be no horror experienced or undergone.

My death can be bad for other people, and for me to imagine or envision my own death is emotionally distressing. But—the Epicureans maintained—even if I were to die at noon tomorrow, it would not be bad for me; I would not be suffering any deprivation. There would be no more me. I would no longer exist to experience anything, including pain or distress or anxiety concerning my loved ones. For something to be good or bad for a being is for it to be conducive to pleasure or pain, and there is no possibility of either once I am dead. Nor is the process of dying to be greatly feared. Intense pains, Epicurus thought, can only last a short time, and any pain that lasts a long time is mild. Death is likely to be a not very painful process in which the capacity to experience and so to feel pain diminishes as death approaches.

It is questionable whether these arguments can actually diminish the fear of death and the aversion to it. Contrary to Epicurus' assertion, dying, when strong drugs are not at hand, can be intensely painful. Further, Nagel's deprivation argument captures what is bad about a death that arrives unwished for. Suppose I am dismayed at not being invited to what promises to be a delightful party I know to be taking place tomorrow. Nothing, it seems, can remove my annoyance short of actually receiving a last-minute invitation or discovering that the party will not be delightful after all.

In the same way, I can be dismayed at the thought of missing the delightful future that is involved in merely being alive. Nothing can remove my dismay except an intervention that saves my life, or the discovery that life after a certain point will not be delightful after all. My dismay at the prospect of my eventual death, probably from natural causes, is milder than my dismay at my impending execution would be, but I can be certain that there will come a time when there will be no intervention that can save me no matter how delightful life continues to be.

The Epicureans deserve credit nevertheless for their courageous stance in trying to remove the awfulness of an inevitability. As noted, against reason and evidence, human beings are strongly motivated to produce and accept reasons for believing in the immortality of the soul. This is a manifestation of our love of life and experience, and our attachment to friends and family. We miss and grieve for people whose presence and conversation are overwhelmingly important for us. We wish to continue to delight in their presence, to follow their lives, to offer our companionship and existence. The thought that after some particular day in some particular year I shall never hear my child's voice, or see a full moon again, or enjoy my morning coffee or the scents of spring is not easy to bear. But if materialist philosophy is in the end powerless fully to prepare or console us, this is not surprising in light of its own claim that that we are material beings whose terrestrial experiences are all that are truly important to us.

Epicurean ethics and human welfare

Despite strong opposition from critics who repeatedly described the Epicureans as greedy swine and depraved libertines, their defence of pleasure and their refusal to ennoble suffering had an important influence on ethics and political philosophy.

As industry, trade, and wealth began to develop in 18th-century Europe, and the living standards of the middle classes began to rise, Epicureanism began to be associated with luxurious city living. The ideals of Naudon and his plans for an Epicurean shopping centre would have not have seemed at all strange to a certain sector of the emerging bourgeois society. Good cooking, the arts of gardening and furnishing houses, moving plays and hilarious comedies, intricate music, and kindly and tender manners in circles of intimates were all valued.

The pursuit of luxury (though commended by neither Epicurus nor Lucretius), Bernard de Mandeville argued, keeps the wheels of the economy turning and feeds the poor. The secularized moral theory of David Hume and Adam Smith laid weight on the pleasure-inducing and -enhancing features of right conduct for the moral agent and his fellows, and on the importance of social approval and disapproval in shaping the mores of a given society.

At the same time, the growth of the new commercial state was correlated with the misery of the worker that Marx would document. Here, the Epicurean outlook was relevant in a different way. The focus on pleasure and pain, the doctrine of mortalism, and the interest in the evolution of culture fed into the ideals of democracy and equality. One life is all we have, distinctions of status are merely the product of the imagination and the contingencies of history, and life ought to be made as trouble free for all as possible. The poor will not receive anything better in heaven. The notion of the 'general welfare' that appears in utilitarian political theory is centred not on the pursuit of pleasure by those who can afford it, but on the relief of pain and deprivation for the anonymous masses. The ideal of equality commands a pleasant state of existence for all, not just for the privileged few.

These notions gathered force in the late 18th century, and furnished the conceptual basis for the impressive utilitarian reforms of the 19th aimed at reducing poverty and disease, mitigating the cruelty of punishments, and correcting abuses of the legal system and corrupt practices. Jeremy Bentham, the founder, along with John Stuart Mill, of utilitarianism, echoed Epicurus in declaring at the start of his influential treatise *The Principles of Morals and Legislation* of 1789, that 'Nature has placed mankind under the governance of two sovereign masters, *pain* and *pleasure*. It is for them alone to point out what we ought to do, as well as to determine what we shall do.'

Modern theorizing about happiness partially endorses the Epicurean position. Daniel Kahneman suggests that a happy life involves both enjoyable episodes such as dining out with friends, loafing on a beach, and pursuing a hobby, but also significant experiences that one can return to in memory—a point also stressed by Epicurus, who noted that we do not simply live in the present but in our anticipation of the future and in our recollections.

In this regard, Epicurus thought, humans are superior to other animals. They have the power to foresee the future and to remember the past, and they have the ability to make choices accordingly. Making certain choices now can relieve us from fear of the future and later result in gratitude for our pasts. The memory of past goods we have enjoyed can console us for those we lack now. Confirming the Epicurean emphasis on friendships based on liking rather than alliances based on necessity or advantage, W. J. Freeman cites neurological evidence that the deepest forms of happiness arise from cooperative activities undertaken with others that evoke feelings of trust and security.

The Epicurean valorization of experiences and shared pleasures seemed to Epicurus' critics to miss the point where ethics was concerned. They did not dispute the premise that all animals shun pain and seek pleasure. However, they—the Stoics in particular—went on to insist that humans have powers and responsibilities far transcending those of the animals in light of their rationality. In ethics, they insisted that the *honestum*—the right—was different from and opposed to the pleasant, and that human beings had an innate capacity to recognize the right and the good and to pursue it. Christianity joined the ideal of Stoic self-command and strength of character to mortification, elevating poverty and the ascetic life, and even martyrdom—the emulation of Christ on the cross—to ethical ideals. In neither Stoicism nor in Christianity was happiness identified with pleasure, or securing happiness considered the goal of ethics.

Four specific objections seem to have some force against the Epicurean answers to the question how we should live. First, there are serious arguments against all forms of eudaemonism, in which securing personal happiness is the aim of ethics, whether or not happiness is identified with pleasure. Kant, for example, declared that the study of how to be happy is an entirely different subject from the study of how to behave in a morally upright fashion. The former depends on skill and luck and is not fully under our control, he thought, while the latter depends only on cultivating a good will, which any rational being can do.

As observed, the Epicureans appear to recognize and respect the virtues of truthfulness, loyalty, courage, and altruism, but only insofar as their practice makes life more pleasant and prevents painful episodes. But, Kant argued, what if this relationship fails to hold, and the exercise of the virtues fails to produce pleasure and reduce pain? Where ethics is concerned, it is not a matter of performing social experiments to see how far we can enjoy ourselves without getting into trouble, but a matter of acting as reason dictates, regardless of the predicted consequences. This position seems to have convinced many moral philosophers.

Second, adopting the goal of minimizing pain and maximizing long-term though not short-term pleasures might have the paradoxical consequence that we must welcome pain. As Socrates noted, an itch is the condition of the pleasure of scratching. The end of any experienced pleasure involving 'movement' (so-called 'katastematic' pleasure) must be somewhat painful. Leibniz too perceived that pleasure must be preceded by 'unease' to be felt as pleasure, and that a true equilibrium of constant pleasure is impossible to achieve.

Other, more recent criticisms of pleasure as a goal invoke the notion of the 'hedonic treadmill' and the 'set point'. As existing desires are satisfied, new ones arise, so that one must keep exerting oneself to escape a painful state of deprivation. Further, modern psychological research indicates that people's hedonic levels have set points, so that neither misfortunes such as losing a limb nor unexpected good fortune such as winning the lottery make any difference to reported happiness after a year or so has gone by. This suggests that whether one has a pleasant or unpleasant life depends less on 'choice and avoidance' and more on how one was constituted by nature in the first place.

Third, some admitted virtues and vices, including the virtue of social concern, pertain to acts or omissions that are beyond the scope of reciprocity and retaliation. Why should I personally expend resources, time, money, or thought on the plight of distant strangers, or donate money to relief organizations working in Central Africa? They cannot retaliate against me for my neglect, and my conscience may not trouble me in the least. The Epicurean can give no plausible answer to this question.

Finally, to aim to minimize pain, one must aim to minimize the risk of pain and the potential for regret. But a risk-averse strategy can cut us off at the same time from the greatest pleasures in life. Epicurus is surely right to say that by foregoing marriage and children one can avoid grievous outcomes, such as troublesome or ungrateful children or the death of a beloved spouse. Observation and sober calculation may indicate that the likelihood of such adverse events is high. Why not give it a go anyway? The Epicurean position—that one can only be in a deprived state if one feels oneself to be in such a state—runs up against our intuitions, not only about mortality but about the possibility of being blind to what we would enjoy if we only had it or had chosen it rather than avoided it at some earlier time.

Chapter 9
The Epicurean legacy

In this short book, I have tried to present Epicureanism in as
sympathetic a light as possible. I have tried to show how the
atomic theory has been adopted, challenged, and transformed
since its introduction, and how the original Epicurean notions of
chance, natural motion, and 'limits' became the basis of the
mechanical philosophy of the Scientific Revolution. I looked at
the theory of many worlds, at the preference for physicalist
explanations of all phenomena, and at the ingenious but
somewhat problematic accounts of perception and thinking
offered by Epicurus and Lucretius.

I went on to consider the atheism of the Epicureans and their
account of the natural origins of human beings and other animals,
and noted their influence on later evolutionary theory and on
theories of law, morality, and justice. I considered their account
of the evolution of society and the way in which it furnished a
basis for both pessimistic and optimistic assessments of where
civilization is headed. I concluded with their treatment of pleasure
and pain, their recommendations for living in a world in which
the gods have never taken an interest, and their attitudes towards
life, love, and death.

Atomism is important as a metaphysics in that it distinguishes
sharply between the world experienced by sentient creatures

and the world as described by scientific theory. For the ancient Epicureans, the atoms, the ultimate constituents of mind-independent reality, were devoid of most of the characteristics of 'solid objects'. They possessed neither colour, nor taste, nor scent, only weight, shape, and hardness. The scents, tastes, and appearances of solid objects depended, the ancient Epicureans maintained, on the body and mind that are affected by them. Thus, different species of animal experience different worlds according to their own material constitutions, and different people, and people in different conditions of health and disease, will also inhabit qualitatively different worlds.

The 17th-century corpuscularian theory of ultimate reality was a version of atomism that steered clear of controversial commitments to the void, the indivisibility of the primary particles, and material minds or 'thinking matter'. This theory introduced the Christian God in a new role as the author of the newly discovered mathematized laws of nature and provided a framework for the new experimental science of the late 17th century. It was of crucial importance for the evolution of scientific method as well as the conceptualization of physics, chemistry, psychology, and medicine.

From the perspective of our contemporary physics, the reduction of the visible world to an appearance founded upon minute material corpuscles was on the right track. Images of molecules, such as that depicted in Figure 9, and even individual atoms of particular elements, can be seen and captured with modern quantum and electron microscopes.

Nevertheless, the most basic entities posited by contemporary physics do not possess definite shapes and fixed weights, and their motion and location cannot be assigned with precision. It may also be false to say of these entities that they cannot undergo any change on the grounds that all change implies the existence of parts and the ultimate constituents of reality must be simple. For

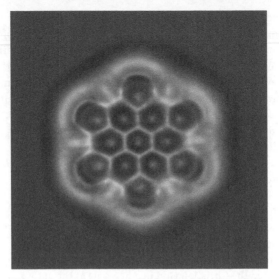

9. A hexabenzocoronene molecule produced with an atomic force microscope.

the solid objects we know, this is the case, but the very notions of 'parts' and 'wholes', along with 'velocity', 'distance', and 'proximity', may not apply to the fundamental particles discovered by physics.

For some theorists, such as David Bohm, the particles located in space-time that are the object of empirical study belong to an 'explicate order', while the 'implicate order' that gives rise to them cannot be detected by observers; its causal relationship to the explicate order cannot accordingly be explained. This view is related to Kant's doctrine of the non-material 'thing in itself' that constituted for him a metaphysically preferable alternative to the Epicurean atom.

Turning to the theory of perception, the Epicurean view that the qualities of the world depend on the constitution of the experiencing creature, and that all animals with sense organs perceive, is clearly correct. The Cartesian position that only human beings have experiences is agreed by nearly all researchers to be an unwarranted assumption. We should have no doubt that foxes and eagles have experiences of objects and situations, though we do not know and cannot know what their experiences are like.

Man is not the measure of all things. Nevertheless, we experience a sensory world that is characteristically human, enabling us to communicate and cooperate with others, and that is correlated to our habits, actions, and capabilities. As phenomenologists have stressed, this adaptation proceeds down to the level of individual differences: photographers, musicians, mathematicians, and schizophrenics all have different visual, auditory, and tactile experiences, and these correspond to differences in the anatomy and physiological workings of the individual's brain, as the materialist supposes. The Epicurean is always interested in understanding *why* people believe the things they do, even when their beliefs seem unwarranted or harmful, rather than simply dismissing them as wrong and irrational. This can be seen in Epicurus' treatment of religion.

The Epicurean insistence that perception is the criterion of all truth seems at first to fit poorly with the notion that the world as we experience it is entirely different from the world which causes our experiences. It is also difficult to reconcile with the notion that experienced worlds are relative to perceivers of different species and different types. It is hard to see how different assessments of the qualities of the same objects, events, or situations can all be equally true, and Epicurean epistemology has been considered incoherent on this account since ancient times.

Accepting the claim that perception is the criterion of truth should not, however, lead me to suppose that train tracks really meet in a point down the line, or that the sun is sometimes the size of a dinner plate (though Epicurus seemed to think it was), and to adjust my behaviour accordingly (by, for example, not taking trains or rejecting the declarations of astronomers regarding the size of the sun). Other perceptions—that trains usually get to where they are going, or that, according to books whose testimony people accept, well-trained astronomers have conducted measurements and made inferences establishing the size of the sun—can be preferred to my immediate ones.

The Epicurean position that one ought to look for explanations that cite material causes rather than abstract entities or occult powers has much to recommend it. Rather than ascribing someone's behaviour to an 'inferiority complex', or some other poorly understood or fictional entity, for example, we might look to concrete experiential, physical, and environmental causes. At the same time, the Epicurean view that just such a causal explanation of any striking phenomenon is worth seeking because it removes fear, especially fear of wrathful divinities, is foreign to the contemporary understanding of why scientific enquiry is valuable. Scientific explanations can be sought, we suppose, for the intrinsic satisfaction of knowing how things work, even if the explanation produces no utility and removes no fear. In this regard, Aristotle and the Stoics perhaps had a better grasp of the intrinsic satisfaction of intellectual understanding.

The Epicurean readiness to entertain a multiplicity of moderately plausible explanations for some phenomenon so long as no single one can be singled out as authoritative is of obvious use in fields like medicine, where the precise mechanisms behind the most common afflictions of modern humans are still poorly understood. Ultimately, however, we would like to know, insofar as prediction or well-intentioned intervention on the basis of a wrong theory can have disastrous consequences, whereas prediction and intervention on the basis of a correct theory can produce gratifying results. The invention of scientific instruments, experimental protocols, and techniques of measurement and analysis have made it possible to arrive—sometimes—at explanations we can accept as 'empirically adequate' and sufficiently well established to serve as just such a basis. It is nevertheless important to remember that the world beneath and beyond the life-world of immediate experience may not be perfectly knowable.

The Epicurean teaching on the gods, as shown in Chapter 6, is somewhat difficult to pin down. The gods are sometimes interpreted as dream images or constructions of the imagination that enter into and reside in the mind, sometimes as mind-independent 'blessed' living beings that inhabit remote regions of the universe where they ignore us. Some may feel that the Epicureans—Lucretius especially—failed to recognize the benefits of belief in an all-powerful caring God and a ruling Providence. Some contemporary studies suggest that theists are happier than atheists. Participation in rites and ceremonies deemed superstitious by the Epicureans provides for many people an opportunity for moral reflection and a sense of fellowship with others, as well as an outlet for charitable activities. The superb art, architecture, and music of the Renaissance and Baroque periods were produced in a religious context.

The Epicurean atheist may insist that such opportunities and such creativity can manifest themselves in secular institutions and from secular motivations just as well. However, we do not know whether, to stimulate the ethical commitment and artistic creativity of which human beings are capable, they can do without a sense of the numinous that is in some way personalized. For Lucretius, it might be pointed out, nature—which does provide him with just such a sense of the numinous—is personalized as the goddess Venus.

The Epicureans were surely right to identify and criticize the fear- and violence-inducing aspects of many organized religions. The notions of sacrifice, martyrdom, punishment, and holy war are central to Judaeo-Christian-Islamic monotheism. The Old Testament emphasizes the vengeful nature of Jehovah, who drives Adam and Eve from Paradise, destroys entire cities, and sends plagues to punish his own supposedly beloved people for sin and disobedience. The doctrine of hell as a place of fiery torment was vividly presented in paintings and sermons and its geography articulated by Dante in the *Divine Comedy*. Many a child has been cruelly subjected to and terrified by tales of the Devil and threats of eternal damnation.

Would an atheistic world actually be more pacific, or would human beings organize equally bloody conflicts around secular issues such as language, customs, trade, and territory? We do not know the answer to these questions, but surely widespread acceptance of the view that morality is not a set of divine commandments, but a set of conventions for peaceful coexistence and human flourishing, would put an end to much acrimonious debate and the persecution of the many individuals who are viewed as having broken a god's laws.

In the late 18th century, the Marquis de Sade expounded an anti-morality in which pleasure—understood as the psychological pleasure of dominating others and the sensual pleasure of extreme thrills—was the supreme good and was to be pursued at all costs, even if it inflicted torture and death on other persons. Embedded as it was in an atheistic philosophy of nature, Sade's philosophy could be seen as Epicureanism taken to its logical conclusion. This would be a mistake. The infliction of pain on one's intimate friends or on strangers is contrary to Epicurean moral teaching, as is the immoderate pursuit of sensual experience.

Nevertheless, Epicurus' refusal to acknowledge an absolute standard of justice, and his provision of exclusively self-interested motives for being moral, have long seemed to some philosophers, as they seemed to the Stoics, to miss the point of morality. Kant faced a problem in this regard that he refers to at several points in his writings. The Stoics, Kant believed, could supply what was essentially a correct theory of morality in their account of the virtues, but they furnished no incentives for engaging in moral behaviour. The Epicureans had no difficulty in supplying incentives, including the expectation of reciprocity, the fear of punishment, and the gnawings of conscience, but these motives were impure in Kant's view (insofar as happiness was the goal) and unreliable. Kant decided that as long as there was the metaphysical possibility of a 'pure' moral motivation to do the right thing, residing in a 'noumenal' subject unbound by either the laws of nature or divine command, we should believe in it.

The Epicurean moralist might well agree with the Kantian that it is morally right to forbear from harming another, even when it is in one's long-term self-interest to do so, that we have a deeply rooted feeling of respect for justice and injustice, and that ethics involves choices. These convictions do not, however, point to anything transcendental: only our beliefs, desires, and social experience. Our motivation can only be directed towards the particular codes we have been educated to heed, and, as circumstances and our understanding of them changes, our motivations to obey or disobey will change as well. We renounce many opportunities for harm because we count on others to do so as well, and perceive the benefits of mutual forbearance. An Epicurean might well add that the moral sentiments are experienced as implacable commands in a manner analogous to the way in which certain mental images of stately beings are interpreted as experiences of the gods.

In Lucretius' poetry, compassion for the weak and suffering appears as an important moral and philosophical concept that is otherwise lacking in ancient philosophy, in which the weak are generally objects of contempt. The utilitarian, welfare-oriented outlook on justice of the late 18th and 19th centuries, the conviction that all are equally deserving of agreeable conditions and pleasant experiences, and the attempts to remedy the appalling social conditions of early capitalism through prison reform, sanitation, protection and education of the labourer, and later through the redistributive policies of the welfare state, reflected a fading belief in the powers of Providence and the rediscovery of Epicurean ethics.

Although Bentham's utilitarian successor, John Stuart Mill, found Bentham's deliberate refusal to distinguish between 'higher' and 'lower' pleasures unacceptable (Bentham notoriously declared that push-pin was as good as poetry), he insisted that the authentic Epicurean tradition assigned a much higher value to 'pleasures of the intellect, of the feelings and imagination, and of the moral sentiments', than to pleasures of 'mere sensation'.

This may be an exaggeration on the part of this high-minded Victorian author; Epicurus himself refers to the 'cry of the flesh' as supremely important. The foregoing will have made it clear, however, that the Epicurean advice for living is not directed towards fancy cooking, luxurious clothing, overindulgence in drink, sexual promiscuity, ambitious or conquistadorial conduct, or other forms of sensation-seeking and self-indulgence, but rather towards enjoying the harmless pleasures life has to offer while avoiding the pains of deprivation and emotional disturbance. Where their advice for daily living is concerned, both the Stoics and the Epicureans valued tranquillity. Their understanding of tranquillity and its attainment was divergent, however.

For the Stoics, participation in political and family life, including the raising of children, was a duty. Both forms of participation, they recognized, created stresses and opportunities for humiliation, loss, and psychological pain, in addition to the physical pains of illness and the unavoidable discomforts of old age. Political participation entailed such evils as the loss of power and the seizure of control of government by one's enemies or by detested ideologues or tyrants; it could result in execution, banishment, or forced suicide. Love of spouse and children and loyalty to family exposed one to loss.

It was necessary, the Stoics believed, to adopt a philosophical attitude towards the unfurling of events, understanding their inevitability, and fortifying oneself against painful emotions, registered in the body as heartache, anxiety, fear, and depression.

The key idea was that the mind has considerable control, not over what happens, but about how one feels about what happens.

There is beauty and nobility in the Stoic conception of *amor fati*, love of fate, and a world ruled by Providence in which we must accept and indeed embrace our destinies, with all the good and bad that arrives. Many people have found some degree of solace for their political and personal misfortunes in Stoic philosophy. The Epicureans, by contrast, had no such exalted notion of the ultimate wisdom of Providence and no motive to love of fate. Nor were they optimistic about the mind's power over the emotions, insofar as the mind, in their view, could not help but register almost everything important that happened to the body or in it.

In their view, the world is undetermined and unpredictable, not only on account of the famous 'swerve', but because every moment presents us with new configurations, and nature is constantly innovating. We have only partial control over external events, which depend on nature and the actions of other people. And, as material beings whose soul atoms are intermingled with our flesh, we have very limited control over our emotional reaction to these events.

Furthermore, we have only partial control over our own thoughts, and we are at the mercy of our turbulent physiological reactions to the material images that invade our minds from outside. Hence the Epicureans assign little importance to the psychological strategies to be employed when confronted with personal upsets and tragedies, and more to the strategies for living that either reduce the probability of getting into an emotionally troubling situation in the first place, or that promptly show us the way to the nearest exit.

For the Epicureans, tranquillity was to be achieved first by shaking off frightening superstitions, then by avoiding overly demanding social practices. If misfortunes nevertheless occur, or illness strikes us down, some consolation can be found in the thought that pains are either intense but short-lived or long-lived but mild. As long as we are suffering, we are alive, and when we cease to be alive, we will also cease to suffer.

If one of the strengths of Epicurean moral philosophy is the role it allows for compassion and mutual agreement, one of its weaknesses is its 'selfishness' and aversion to risk. Again, Epicurean selfishness is not to be confused with egoism; the Epicurean precepts are against self-aggrandizement. But as tempting as it is to leave politics to others, it is morally difficult in the modern world to 'live apart'. Our habits of consumption and voting practices affect others in harmful ways, and leaving politics to the quarrelsome and ambitious has, it could be argued, produced a world increasingly dangerous for us and others. And is it really advisable to avoid family life and love on the grounds that they offer too many opportunities for disappointment?

George Ainslie, in one of the most interesting recent treatments of human motivation, writes in favour of the pleasure of immediate, personal experience but against the ideal of tranquillity, whether Epicurean or Stoic. He observes that:

> [i]f you subtract the relief of physical hungers from the set of strongly motivated human goals, what is left are events that 'induce' emotion. The great novels are about romantic love and love of family, the desire for glory and the desire for revenge, the gratification that comes of dominating others and the resentment, or satisfaction, that comes of being dominated.

Thus, he concludes, 'the greater part of wealth should perhaps be equated with the prospect of emotional experience', and this, for Ainslie, involves the acceptance of risk, the need 'to gamble enough that you lose sufficiently often'. To the extent that moral philosophies advise us to dampen emotion or limit exposure to emotion-generating situations, to replace uncertainties with certainties, and openness to the unpredictable with control, they may have missed this important component of human nature. The pleasure of reading fiction arises from the vicarious experience of emotion and risk-taking it offers in conditions of personal security, but it does not altogether replace the real.

To conclude, the Epicurean depiction of the world of solid objects as a lively, colourful, meaningful appearance, founded upon what its 17th-century opponents referred to as the 'random concourse of atoms' possessing only magnitude, shape, and motion, was both inspiration for and anathema to generations of later philosophers and their readers. The moral, political, and theological ideas associated with it stimulated metaphysics, science, and social thought, arousing at times furious resistance.

Lucretius described Epicureanism as a 'bitter medicine' he hoped to sweeten with poetry, and he succeeded in making a case for its saving power without, however, avoiding its pessimism. One comes away from *On the Nature of Things* with admiration for the comprehensiveness of the Epicurean theory of nature and an appreciation not only of the limits to life, power, and enjoyment, but also of the unlimited ability of nature to renew and create, an ability ensured by the indestructibility and unceasing motion of the *primordia*—whatever these turn out to be—of nature.

Epicureanism is not a fatalistic philosophy. It lays great weight on human choices and preferences and on the role of chance—the atomic 'swerve'—in presenting challenges and opportunities that call on our powers of invention as well as our rationality. It is the most compassionate of the ancient ethical systems. It invites us to take pleasure in what is near at hand: in warmth, food, and drink, in moderation; in the company of those we happen, for whatever reason, to like; in the recurrence of spring after winter; and in the surround of foliage and flowers, and the appearance of new life.

Publisher's acknowledgements

We are grateful for permission to include the following copyright material in this book.

Excerpts from *Discourse on the Origins of Inequality (Second Discourse), Polemics, and Political Economy*, edited by Roger D. Masters and Christopher Kelly, published by Dartmouth College Press. Used with permission.

The publisher and author have made every effort to trace and contact all copyright holders before publication. If notified, the publisher will be pleased to rectify any errors or omissions at the earliest opportunity.

References

The quotations from Epicurus in the text are taken from the following
sources: *The Epicurus Reader* (ER), edited by Brad Inwood and
L. P. Gerson, with an introduction by D. S. Hutchinson (Indianapolis:
Hackett, 1994).

Lucretius' *On the Nature of Thing* (NT) is quoted after the translation
by Martin Ferguson Smith (Indianapolis: Hackett, 1994).

The Greek-English Loeb edition of Diogenes Laertius (DL), *Lives of
Eminent Philosophers*, tr. R. D. Hicks (Cambridge, MA: Harvard
University Press, 1931), is widely available. Volume II, Ch X is
devoted to Epicurus. The Latin-English version of Lucretius' *De
Rerum Natura*, tr. W. H. D. Rouse, revised by Martin Ferguson
Smith (Cambridge, MA: Harvard University Press, 1992), belongs
to the same series.

Chapter 1: Introduction

'"This is where the butcher's shop will go,"...silk scarf and a Rolex', Liz
Alderman, 'An Epicurean Village Is Too Rich for Some Paris
Appetites,' *New York Times*, 9 August 2014.

'modern holders of the doctrine...and English assailants', J. S. Mill,
Utilitarianism, On Liberty and Other Essays, ed. John Gray
(Oxford: Oxford World's Classics, 1991), p. 138.

'I know not how to conceive the good...pleasures of beautiful form'
(DL X: 6).

'unsurpassed goodwill...affection for his country' (DL X: 10).

Chapter 2: Atomic worlds

dust motes dancing in a sunbeam (NT II: 112–41).

'Consider the iridescence…with blue lazuli' (NT II: 801–9).

'motionless pool of bright light upon the plain' (NT IV: 317–33).

an alphabet analogy (NT II: 688–99).

For the beautiful passages of Cicero's treatise *On the Nature of the Gods*, see especially Bk II, Chs XI–LXIV reporting the views of the Stoic Chrysippus, in *De Natura Deorum* and *Academica*, tr. H. Rackham (Cambridge, MA: Harvard University Press, 1951) pp. 151–279.

John Locke distinguished between the 'primary' properties of matter (size, shape, motion, weight, and solidity). See John Locke, *An Essay Concerning Human Understanding*, ed. P. H. Nidditch (Oxford: Clarendon Press, 1975), II.viii.9–106.

'held in union by…hooks and eyes' (NT VI: 1087–8).

Chapter 3: Knowledge and understanding

'[t]he conviction that comets…"against the stars"', George Johnson, 'Comets Breed Fear, Fascination and Web Sites,' *New York Times*, 28 March 1997.

'clings always…similar to the phenomena' (ER: 24).

'Our earth and sky…disease to be produced' (NT VI: 663–4).

'on account…component atoms' (NT VI: 775–6).

'the method of unique explanations…what cannot be understood' (ER: 21).

'Human knowledge and human power meet in one', Francis Bacon, *Works*, ed. J. Spedding, R. L. Ellis, and D. D. Heath, 15 vols (Boston: Houghton Mifflin, *c*.1900) VIII: 67.

Chapter 4: Living, loving, dying

'seeds of wind…quits the bones' (NT III: 120–9).

'[A]ll of us are sprung…reproduce their kind' (NT II: 991–7).

'living creatures cannot have dropped from heaven…gulfs of the sea' (NT V: 793–4).

'the old Epicurean hypothesis…yet been proposed', 'might not… appearance of probability', and 'A finite number of particles…an infinite number of times', David Hume, *Dialogues Concerning*

Natural Religion, and the Natural History of Religion, ed.
J. C. A. Gaskin (Oxford: Oxford University Press, 2008), pp. 84–5.

The seminal fluids of male and female mingle during intercourse (NT IV: 1210–32).

'Venus, power of life...to reproduce its kind' (NT I: 7–20).

'it is better to marry than to burn [with unsatisfied lust]', 1 Corinthians 7:9.

'[N]o visible object...from hand to hand' (NT I: 261–4, NT II: 75–9).

'It is like the case of a wine...dispersed into the air' (NT III: 221–2).

Chapter 5: Material minds

'the matings and births of wild beasts' (NT III: 776).

A mind requires an animal body (NT III: 784–90).

'very small seeds,...veins, flesh, and sinews' (NT III: 216–18).

'shake and tremble... absolutely devoid of sensation' (NT II: 986–90).

'united and combined' (NT II: 941).

'continually streaming off from the surface of bodies' (DL X: 48).

'outlines...a single continuous thing' (ER: 9).

'saffron, russet, and violet' (NT IV: 75–80).

'countless subtle images...on every side' (NT IV: 724–6).

'we fancy that we pass...plains on foot' (NT IV: 459–60).

We need to grasp what is denoted by our words (ER: 6).

Names are given to things with a general outline or a characteristic appearance (ER: 42).

Square towers look round from a distance (NT IV: 353–9).

The fault...lies not in the senses but in the 'reasoning of the mind' (NT IV: 383–4).

'some slight aid from reasoning' (DL X: 32).

'our conception of truth is derived ultimately from the senses' (NT IV: 479–80).

reasoning...cannot contradict sense (DL X: 32).

'if you were not prepared...at once collapse' (NT IV: 507–8).

the hallucinations of madmen and the dreams of sleepers (DL X: 32).

'little images flitting though the air', Rene Descartes, *Optics*, in *Philosophical Writings of Descartes*, tr. J. Cottingham, R. Stoothoff, and D. Murdoch, 2 vols (Cambridge: Cambridge University Press, 1984–5), I: 153–4.

Chapter 6: Religion and superstition

'because nature herself...minds of mankind' (Cicero, *Nature of the Gods*, I: XVI, 45).

all the minds endowed with reason and purpose that we know (Cicero, *Nature of the Gods*, I: XXXI, 84).

this does not seem to be what Epicurus had in mind (ER: 52–3).

'the first who dared to lift mortal eyes...grinding weight of superstition' (NT I: 62–3).

'The minds of mortals...stupendous stature' (NT V: 1170–2).

'the orderly movements...seasons of the year' (NT V: 1183–6).

'O hapless humanity...generations to come!' (NT V: 1194–9).

'neither the worship of the gods, nor their divinity counted for much' (NT VI 1275).

'The time may come...punishment without end' (NT I: 102–11).

'ignorance of natural causes...unintelligible opinions', Paul-Henri Thiry, Baron d'Holbach, *Good Sense* (1772), trans. L Stewart (London: 1900), preface.

'fortuitous concourse of atoms', Cicero, *Nature of the Gods*, II: XXXVII; 93.

'[a]ll these things...universe is the assemblage'(Paul Henri Thiry, Baron d'Holbach, *Good Sense*, tr. Anna Knoop (Amherst, NY: Prometheus, 2004), Sect. 38).

'a sorrowful and sinless victim of a sinful crime' (NT I: 101–2).

'Such heinous acts could superstition prompt' (NT I: 101).

Chapter 7: Politics and society

'the male sex is, generally speaking, far superior in skill and ingenuity' (NT V: 1356–7).

'not knowing the limits of possession' (NT V: 1423–4).

'gold and purple...plague human lives with cares and weary them with war' (NT V: 1412–32).

'always moving, sweating, toiling,...laborious occupations' and 'more violent outrages...on the whole earth', Jean-Jacques Rousseau, *Discourse on the Origins of Inequality*, *Collected Writings of Rousseau*, 13 vols, ed. Roger D. Masters and Christopher Kelly, tr. Judith R. Bush, Roger D. Masters, Christopher Kelly, Terence Marshall, and Allan Bloom (Hanover and London: University Press of New England, 1990–2010), III: 78.

Chapter 8: Epicurean ethics

both mind and body succumb to the 'stress and strain of age' (NT III: 458).

'all good and bad consists in sense-experience' (ER: 29).

'searches out the reasons for every choice and avoidance' (ER: 30).

'It is not drinking bouts...but sober calculation' (ER: 31).

'drops of water falling upon a rock in long lapse of time' (NT IV: 1286-7).

'beckoned by divine pleasure' (NT II: 173).

'honeyed drop of Venus' sweetness...followed by chilling care' (NT IV: 1059-60).

'ambiguous word'...or the 'trace of a smile' (NT IV: 1137-41).

'we have lived without her until now' (NT IV: 1173-4).

'random-roving' Venuses you don't love (NT IV: 1072).

'When we exist...then we do not exist' (DL X: 125).

'Nature has placed mankind...determine what we shall do' Jeremy Bentham, *An Introduction to the Principles of Morals and Legislation* (New York: Dover, 2007), p. 1.

The memory of past goods we have enjoyed can console us for those we lack now (ER: 101).

Chapter 9: The Epicurean legacy

'pleasures of the intellect...mere sensation', J. S. Mill, *Utilitarianism*, ed. Roger Crisp (Oxford: Oxford University Press, 1998), p. 17.

'[i]f you subtract the relief...comes of being dominated' and 'the greater part of wealth...emotional experience', George Ainslie, 'Uncertainty as Wealth,' *Behavioural Processes*, 64(3) (2003), pp. 369-85: 371.

'to gamble enough that you lose sufficiently often', Ainslie, 'Uncertainty as Wealth,' p. 381.

W.J. Freeman, 'Happiness Doesn't Come in Bottles,' *Journal of Consciousness Studies*, 4 (1997), pp. 67-70.

Further reading

General overviews of Epicureanism include Howard Jones, *The Epicurean Tradition* (London: Routledge, 1989) and David Konstan, *A Life Worthy of the Gods: The Materialist Psychology of Epicurus* (Las Vegas: Parmenides Publishing, 2008). David Sedley in *Lucretius and the Transmission of Greek Wisdom* (Cambridge: Cambridge University Press, 1998) discusses Lucretius' transformation of Epicurus' works, as does Diskin Clay, *Lucretius and Epicurus* (Ithaca, NY: Cornell University Press, 1983). Neven Leddy and Avi S. Lifschitz (eds.), *Epicurus in the Enlightenment* (Oxford: Voltaire Foundation, 2009), present a collection of essays on Epicureanism in the 18th century. For Lucretius' influence, see Stephen Greenblatt, *The Swerve* (New York: Norton, 2011); Alison Brown, *The Return of Lucretius to Renaissance Florence* (Cambridge, MA: Harvard University Press, 2010); and Catherine Wilson, *Epicureanism at the Origins of Modernity* (Oxford: Clarendon Press, 2008).

On specific Epicurean and Lucretian themes, and the social and political context of the two philosophers, two volumes from the Cambridge University Press Companion series are highly recommended: *The Cambridge Companion to Epicurus* (2009) edited by James Warren, and *The Cambridge Companion to Lucretius* (2007) edited by Stuart Gillespie and Philip Hardie. Elizabeth Asmis in 'Lucretius' New World Order: Making a Pact with Nature', *The Classical Quarterly*, 58 (2008, pp. 141–57), relates Epicurean physics to ethics with an emphasis on the important notion of 'limits'. A fascinating but controversial essay by Benjamin Farrington, 'The Gods of Epicurus and the Roman State', *The*

Modern Quarterly, 1 (1938), pp. 214–32, argues that Epicureanism was a politically subversive popular philosophy. Martha Nussbaum in 'Therapeutic Arguments: Epicurus and Aristotle', in *The Norms of Nature*, ed. M. Schofield and G. Striker (Cambridge: Cambridge University Press, 2007), pp. 31–74, discusses the function of philosophical argument as a guide for the perplexed and a help for the distressed.

On specific topics, the following are recommended.

On atomism: David J. Furley, *The Greek Cosmologists* (Cambridge: Cambridge University Press, 2006); Monte Johnson and Catherine Wilson, 'Lucretius and the History of Science', in *The Cambridge Companion to Lucretius*, pp. 131–48. For particles (and fields) in contemporary physics, see John Polkinghorne *Quantum Theory: A Very Short Introduction* (Oxford: Oxford University Press, 2002) and Frank Close, *Particle Physics: A Very Short Introduction* (Oxford: Oxford University Press, 2004).

On the material mind: Stephen Everson, 'Epicurean Psychology', in *The Cambridge History of Hellenistic Philosophy*, ed. Keimpe Algra, Jonathan Barnes, Jaap Mansfeld, and Malcolm Schofield (Cambridge: Cambridge University Press, 2005), pp. 542–59.

On religion and superstition: Benjamin Farrington, *The Faith of Epicurus* (London: Weidenfeld and Nicolson, 1967); and David Sedley, *Creationism and its Ancient Critics* (Berkeley and Los Angeles: University of California Press, 2008). For an updated treatment of religion as a psychological phenomenon, see Pascal Boyer, *Religion Explained: The Human Instincts that Fashion Gods, Spirits and Ancestors* (New York: Vintage, 2002); and Scott Atran, *In Gods We Trust: The Evolutionary Landscape of Religion* (Oxford: Oxford University Press, 2004).

On pleasure: Julia Annas, 'Epicurus on Pleasure and Happiness', *Philosophical Topics*, 15 (1987) pp. 5–21; D.J. Glidden 'Epicurus and the Pleasure Principle,' in D. J. DePew, *The Greeks and the Good Life* (Indianapolis: Hackett, 1980 pp. 177–97). See also Daniel Kahneman, *Well-Being: the Foundations of Hedonic Psychology* (New York: Russell Sage Foundation, 2003) and W.J. Freeman. 'Happiness doesn't Come in Bottles,' *Journal of Consciousness Studies*, 4 (1997), pp. 67–70.

On ethical theory: Philip J. Mitsis, *Epicurus' Ethical Theory* (Ithaca, NY: Cornell University Press, 1988); Suzanne Bobzien, 'Did Epicurus Discover the Free-will Problem?', *Oxford Studies in Ancient Philosophy*, 19 (2000), pp. 287–337.

On perception and knowledge: Elizabeth Asmis, 'Epicurean
Epistemology', in *The Cambridge History of Hellenistic Philosophy*,
pp. 260–94; C. C. W. Taylor, 'All Perceptions are True', in *Doubt
and Dogmatism*, ed. M. Schofield, M. F. Burnyeat, and J. Barnes
(Oxford: Oxford University Press, 1980), pp. 105–24.

On love and marriage: Tad Brennan, 'Epicurus on Sex, Marriage and
Children', *Classical Philology*, 91 (1996), pp. 346–52; William
Fitzgerald, 'Lucretius' Cure for Love in the "De Rerum Natura"',
The Classical World, 78 (1984), pp. 73–86.

On death: James Warren, *Facing Death: Epicurus and his Critics*
(Oxford: Clarendon, 2004); Stephen Rosenbaum, 'How to be Dead
and not Care: A Defense of Epicurus', *American Philosophical
Quarterly*, 23 (1986), pp. 217–25; see also Thomas Nagel, 'Death',
in *Mortal Questions* (Cambridge: Cambridge University Press,
1979), pp. 1–10.

On Epicureans and Stoics: David Furley, 'Lucretius and the
Stoics', *Bulletin of the Institute of Classical Studies*, 17 (1966),
pp. 183–205.

Index

Index

Epicureanism

Expand your collection of
VERY SHORT INTRODUCTIONS